E. C. Russell.
Columbia College.
May, 1876.

PEAT AND ITS USES,

AS

FERTILIZER AND FUEL.

BY

SAMUEL W. JOHNSON, A. M.,

PROFESSOR OF ANALYTICAL AND AGRICULTURAL CHEMISTRY, YALE COLLEGE.

FULLY ILLUSTRATED.

NEW-YORK:

ORANGE JUDD & COMPANY.

245 BROADWAY.

LOVEJOY & SON,
ELECTROTYPERS AND STEREOTYPERS
15 Vandewater street N. Y.

TO MY FATHER,

MY EARLIEST AND BEST

INSTRUCTOR IN RURAL AFFAIRS,

THIS VOLUME

IS GRATEFULLY DEDICATED.

S. W. J.

CONTENTS.

INTRODUCTION.

In the years 1857 and 1858, the writer, in the capacity of Chemist to the State Agricultural Society of Connecticut, was commissioned to make investigations into the agricultural uses of the deposits of peat or swamp muck which are abundant in this State; and, in 1858, he submitted a Report to Henry A. Dyer, Esq., Corresponding Secretary of the Society, embodying his conclusions. In the present work the valuable portions of that Report have been recast, and, with addition of much new matter, form Parts I. and II. The remainder of the book, relating to the preparation and employment of peat for fuel, &c., is now for the first time published, and is intended to give a faithful account of the results of the experience that has been acquired in Europe, during the last twenty-five years, in regard to the important subject of which it treats.

The employment of peat as an amendment and absorbent for agricultural purposes has proved to be of great advantage in New-England farming.

It is not to be doubted, that, as fuel, it will be even more valuable than as a fertilizer. Our peat-beds, while they do not occupy so much territory as to be an impediment and a reproach to our country, as they have been to Ireland, are yet so abundant and so widely distributed — occurring from the Atlantic to the Missouri, along and above the 40th parallel, and appearing on our Eastern Coast at least as far South as North Carolina * — as to present, at numberless points, material, which, sooner or later, will serve us most usefully when other fuel has become scarce and costly.

The high prices which coal and wood have commanded for several years back have directed attention to peat fuel; and, such is the adventurous character of American enterprise, it cannot be

* The great Dismal Swamp is a grand peat bog, and doubtless other of the swamps of the coast, as far south as Florida and the Gulf, are of the same character.

doubted that we shall rapidly develop and improve the machinery for producing it. As has always been the case, we shall waste a vast deal of time and money in contriving machines that violate every principle of mechanism and of economy; but the results of European invention furnish a safe basis from which to set out, and we have among us the genius and the patience that shall work out the perfect method.

It may well be urged that a good degree of caution is advisable in entering upon the peat enterprise. In this country we have exhaustless mines of the best coal, which can be afforded at a very low rate, with which other fuel must compete. In Germany, where the best methods of working peat have originated, fuel is more costly than here; and a universal and intense economy there prevails, of which we, as a people, have no conception.

If, as the Germans themselves admit, the peat question there is still a nice one as regards the test of dollars and cents, it is obvious, that, for a time, we must "hasten slowly." It is circumstances that make peat, and gold as well, remunerative or otherwise; and these must be well considered in each individual case. Peat is the name for a material that varies extremely in its quality, and this quality should be investigated carefully before going to work upon general deductions.

In my account of the various processes for working peat by machinery, such data as I have been able to find have been given as to cost of production. These data are however very imperfect, and not altogether trustworthy, in direct application to American conditions. The cheapness of labor in Europe is an item to our disadvantage in interpreting foreign estimates. I incline to the belief that this is more than offset among us by the quality of our labor, by the energy of our administration, by the efficiency of our overseeing, and, especially, by our greater skill in the adaptation of mechanical appliances. While counselling caution, I also recommend enterprise in developing our resources in this important particular; knowing full well, however, that what I can say in its favor will scarcely add to the impulse already apparent among my countrymen.

SAMUEL W. JOHNSON.

Sheffield Scientific School,
Yale College, June, 1866.

PART I.

THE ORIGIN, VARIETIES, AND CHEMICAL CHARACTERS OF PEAT.

1. *What is Peat?*

By the general term Peat, we understand the organic matter or vegetable soil of bogs, swamps, beaver-meadows and salt-marshes.

It consists of substances that have resulted from the decay of many generations of aquatic or marsh plants, as mosses, sedges, coarse grasses, and a great variety of shrubs, mixed with more or less mineral substances, derived from these plants, or in many cases blown or washed in from the surrounding lands.

2. *The conditions under which Peat is formed.*

In this country the production of Peat from fallen and decaying plants, depends upon the presence of so much water as to cover or saturate the vegetable matters, and thereby hinder the full access of air. Saturation with water also has the effect to maintain the decaying matters at a low temperature, and by these two causes in combination, the process of decay is made to proceed with great slowness, and the solid products of such slow decay, are compounds that themselves resist decay, and hence they accumulate.

In the United States there appears to be nothing like the extensive *moors* or *heaths*, that abound in Ireland, Scotland, the north of England, North Germany, Holland, and the elevated plains of Bavaria, which are mostly level or gently sloping tracts of country, covered with peat or turf to a depth often of 20, and sometimes of 40, or more, feet. In this country it is only in low places, where streams become obstructed and form swamps, or in bays and inlets on salt water, where the flow of the tide furnishes the requisite moisture, that our peat-beds occur. If we go north-east as far as Anticosti, Labrador, or Newfoundland, we find true moors. In these regions have been found a few localities of the *Heather* (*Calluna vulgaris*), which is so conspicuous a plant on the moors of Europe, but which is wanting in the peat-beds of the United States.

In the countries above named, the weather is more uniform than here, the air is more moist, and the excessive heat of our summers is scarcely known. Such is the greater humidity of the atmosphere that the bog-mosses, — the so-called *Sphagnums*, — which have a wonderful avidity for moisture, (hence used for packing plants which require to be kept moist on journeys), are able to keep fresh and in growth during the entire summer. These mosses decay below, and throw out new vegetation above, and thus produce a bog, especially wherever the earth is springy. It is in this way that in those countries, moors and peat-bogs actually grow, increasing in depth and area, from year to year, and raise themselves above the level of the surrounding country.

Prof. Marsh informs the writer that he has seen in Ireland, near the north-west coast, a granite hill, capped with a peat-bed, several feet in thickness. In the Bavarian highlands similar cases have been observed, in localities where the atmosphere and the ground are kept moist enough for the growth of moss by the extraordinary prevalence of fogs. Many of the European moors rise more or less above the level of their borders towards the centre, often to a height of 10 or 20 and sometimes of 30 feet. They are hence known in Germany as *high* moors (*Hochmoore*) to distinguish from the level or dishing *meadow-moors*, (*Wiesenmoore.*) The peat-producing vegetation of the former is chiefly moss and heather, of the latter coarse grasses and sedges.

In Great Britain the reclamation of a moor is usually an expensive operation, for which not only much draining, but actual cutting out and burning of the compact peat is necessary.

The warmth of our summers and the aryness of our atmosphere prevent the accumulation of peat above the highest level of the standing water of our marshes, and so soon as the marshes are well drained, the peat ceases to form, and in most cases the swamp may be easily converted into good meadow land.

Springy hill-sides, which in cooler, moister climates would become moors, here dry up in summer to such an extent that no peat can be formed upon them.

As already observed, our peat is found in low places. In many instances its accumulation began by the obstruction of a stream. To that remarkable creature, the beaver, we owe many of our peat-bogs. These animals, from time immemorial, have built their dams across rivers so as to flood the adjacent forest. In the rich leaf-mold at the water's verge, and in the cool shade of the standing trees, has begun the growth of the sphagnums, sedges, and various purely aquatic plants. These in their annual decay have shortly filled the shallow borders of the stagnating water, and by slow encroachments, going on through many years, they have occupied the deeper portions, aided by the trees, which, perishing, give their fallen branches and trunks, towards completing the work. The trees decay and fall, and become entirely converted into peat; or, as not unfrequently happens, especially in case of resinous woods, preserve their form, and to some extent their soundness.

In a similar manner, ponds and lakes are encroached upon; or, if shallow, entirely filled up by peat deposits. In the Great Forest of Northern New York, the voyager has abundant opportunity to observe the formation of peat-swamps, both as a result of beaver dams, and of the filling of shallow ponds, or the narrowing of level river courses. The formation of peat in water of some depth greatly depends upon the growth of aquatic plants, other than those already mentioned. In our Eastern States the most conspicuous are the Arrow-head, (*Sagittaria*); the Pickerel Weed, (*Pontederia*;) Duck Meat, (*Lemna*;) Pond Weed, (*Potamogeton*;) various *Polygonums*, brothers of Buckwheat and Smart-weed; and especially the Pond Lilies, (*Nymphœa* and *Nuphar*.) The latter grow in water four or five feet deep, their leaves and long stems are thick and fleshy, and their roots, which fill the oozy mud, are often several inches in diameter. Their decaying leaves and stems, and their huge roots, living or dead, accumulate below and gradually raise the bed of the pond. Their living foliage which often covers the water almost completely for acres, becomes a shelter or support for other more delicate aquatic plants and sphagnums, which, creeping out from the shore, may so develop as to form a floating carpet, whereon the leaves of the neighboring wood, and dust scattered by the wind collect, bearing down the mass, which again increases above, or is reproduced until the water is filled to its bottom with vegetable matter.

It is not rare to find in our bogs, patches of moss of considerable area concealing deep water with a treacherous appearance of solidity, as the hunter and botanist have often found to their cost. In countries of more humid atmosphere, they are more common and attain greater dimensions. In Zealand the surfaces of ponds are so frequently covered with floating beds of moss, often stout enough to bear a man, that they have there received a special name "*Hangesak.*" In the Russian Ural, there occur lakes whose floating covers of moss often extend five or six feet above the water, and are so firm that roads are made across them, and forests of large fir-trees find support. These immense accumulations are in fact floating moors, consisting entirely of peat, save the living vegetation at the surface.

Sometimes these floating peat-beds, bearing trees, are separated by winds from their connection with the shore, and become swimming peat islands. In a small lake near Eisenach, in Central Germany, is a swimming island of this sort. Its diameter is 40 rods, and it consists of a felt-like mass of peat, three to five feet in depth, covered above by sphagnums and a great variety of aquatic plants. A few birches and dwarf firs grow in this peat, binding it together by their roots, and when the wind blows, they act as sails, so that the island is constantly moving about upon the lake.

On the Neusiedler lake, in Hungary, is said to float a peat island having an area of six square miles, and on lakes of the high Mexican Plateau are similar islands which, long ago, were converted in fruitful gardens.

3. *The different kinds of Peat.*

Very great differences in the characters of the deposits in our peat-beds are observable. These differences are partly of color, some peats being gray, others red, others again black; the majority, when dry, possess a dark brown-red or snuff color. They also vary remarkably in weight and consistency. Some are compact, destitute of fibres or other traces of the vegetation from which they have been derived, and on drying, shrink greatly and yield tough dense masses which burn readily, and make an excellent fuel. Others again are light and porous, and remain so on drying; these contain intermixed vegetable matter that is but little advanced in the peaty decomposition. Some peats are almost entirely free from mineral matters, and on burning, leave but a few *per cent.* of ash, others contain considerable quantities of lime or iron, in chemical combination, or of sand and clay that have been washed in from the hills adjoining the swamps. As has been observed, the peat of some swamps is mostly derived from mosses, that of others originates largely from grasses; some contain much decayed wood and leaves, others again are free from these.

In the same swamp we usually observe more or less of all these differences. We find the surface peat is light and full of partly decayed vegetation, while below, the deposits are more compact. We commonly can trace distinct strata or layers of peat, which are often very unlike each other in appearance and quality, and in some cases the light and compact layers alternate so that the former are found below the latter.

The light and porous kinds of peat appear in general to be formed in shallow swamps or on the surface of bogs, where there is considerable access of air to the decaying matters, while the compacter, older, riper peats are found at a depth, and seem to have been formed beneath the low water mark, in more complete exclusion of the atmosphere, and under a considerable degree of pressure.

The nature of the vegetation that flourishes in a bog, has much effect on the character of the peat. The peats chiefly derived from mosses that have grown in the full sunlight, have a yellowish-red color in their upper layers, which usually becomes darker as we go down, running through all shades of brown until at a considerable depth it is black. Peats produced principally from grasses are grayish in appearance at the surface, being full of silvery fibres—the skeletons of the blades of grasses and sedges, while below they are commonly black.

Moss peat is more often fibrous in structure, and when dried forms somewhat elastic masses. *Grass peat*, when taken a little below the surface, is commonly destitute of fibres; when wet, is earthy in its look, and dries to dense hard lumps.

Where mosses and grasses have grown together simultaneously in the same swamp, the peat is modified in its characters accordingly. Where, as may happen, grass succeeds moss, or moss succeeds grass, the different layers reveal their origin by their color and texture. At considerable depths, however, where the peat is very old, these differences nearly or entirely disappear.

The geological character of a country is not without influence on the kind of peat. It is only in regions where the rocks are granitic or silicious, where, at least, the surface waters are free or nearly free from lime, that *mosses* make the bulk of the peat.

In limestone districts, peat is chiefly formed from *grasses* and *sedges.*

This is due to the fact that mosses (sphagnums) need little lime for their growth, while the grasses require much;

aquatic grasses cannot, therefore, thrive in pure waters, and in waters containing the requisite proportion of lime, grasses and sedges choke out the moss.

The accidental admixtures of soil often greatly affect the appearance and value of a peat, but on the whole it would appear that its quality is most influenced by the degree of decomposition it has been subjected to.

In meadows and marshes, overflowed by the ocean tides, we have *salt-peat*, formed from Sea-weeds (*Algæ*,) Salt-wort (*Salicornia*,) and a great variety of marine or strand-plants. In its upper portions, salt-peat is coarsely fibrous from the grass roots, and dark-brown in color. At sufficient depth it is black and destitute of fibres.

The fact that peat is fibrous in texture shows that it is of comparatively recent formation, or that the decomposition has been arrested before reaching its later stages. Fibrous peat is found near the surface, and as we dig down into a very deep bed we find almost invariably that the fibrous structure becomes less and less evident until at a certain depth it entirely disappears.

It is not depth simply, but age or advancement in decomposition, which determines these differences of texture.

The "ripest," most perfectly formed peat, that in which the peaty decomposition has reached its last stage,—which, in Germany, is termed *pitchy-peat* or *fat peat*, (*Pechtorf*, *Specktorf*)—is dark-brown or black in color, and comparatively heavy and dense. When moist, it is firm, sticky and coherent almost like clay, may be cut and moulded to any shape. Dried, it becomes hard, and on a cut or burnished surface takes a luster like wax or pitch.

In Holland, West Friesland, Holstein, Denmark and Pomerania, a so-called *mud-peat* (*Schlammtorf*, also *Baggertorf* and *Streichtorf*,) is "fished up" from the bottoms of ponds, as a black mud or paste, which, on drying, becomes hard and dense like the pitchy-peat.

The two varieties of peat last named are those which are most prized as fuel in Europe.

Vitriol peat is peat of any kind impregnated with sulphate of iron (*copperas*,) and sulphate of alumina, (the astringent ingredient of alum.)

Swamp Muck.—In New England, the vegetable remains occurring in swamps, etc., are commonly called *Muck.* In proper English usage, muck is a general term for manure of any sort, and has no special application to the contents of bogs. With us, however, this meaning appears to be quite obsolete, though in our agricultural literature—formerly, more than now, it must be admitted, —the word as applied to the subject of our treatise, has been qualified as *Swamp Muck.*

In Germany, peat of whatever character, is designated by the single word *Torf;* in France it is *Tourbe*, and of the same origin is the word *Turf*, applied to it in Great Britain. With us turf appears never to have had this signification.

Peat, no doubt, is a correct name for the substance which results from the decomposition of vegetable matters under or saturated with water, whatever its appearance or properties. There is, however, with us, an inclination to apply this word particularly to those purer and more compact sorts which are adapted for fuel, while to the lighter, less decomposed or more weathered kinds, and to those which are considerably intermixed with soil or silt, the term muck or swamp muck is given. These distinctions are not, indeed, always observed, and, in fact, so great is the range of variation in the quality of the substance, that it would be impossible to draw a line where muck leaves off and peat begins. Notwithstanding, a rough distinction is better than none, and we shall therefore employ the two terms when any greater clearness of meaning can be thereby conveyed.

It happens, that in New England, the number of small shallow swales, that contain unripe or impure peat, is much greater than that of large and deep bogs. Their contents are therefore more of the "mucky" than of the "peaty" order, and this may partly account for New England usage in regard to these old English words.

By the term muck, some farmers understand leaf-mold (decayed leaves), especially that which collects in low and wet places. When the deposit is deep and saturated with water, it may have all the essential characters of peat. Ripe peat, from such a source is, however, so far as the writer is informed, unknown to any extent in this country. We might distinguish as *leaf-muck* the leaves which have decomposed under or saturated with water, retaining the well established term leaf-mold to designate the dry or drier covering of the soil in a dense forest of deciduous trees.

Salt-mud.—In the marshes, bays, and estuaries along the sea-shore, accumulate large quantities of fine silt, brought down by rivers or deposited from the sea-water, which are more or less mixed with finely divided peat or partly decomposed vegetable matters, derived largely from Sea-weed, and in many cases also with animal remains (mussels and other shell-fish, crabs, and myriads of minute organisms.) This black mud has great value as a fertilizer.

4. *The Chemical Characters and Composition of Peat.*

The process of burning, demonstrates that peat consists of two kinds of substance; one of which, the larger portion, is combustible, and is *organic* or vegetable matter; the other, smaller portion, remaining indestructible by fire is *inorganic matter* or *ash.* We shall consider these separately.

(*a*) *The organic or combustible part of peat* varies considerably in its proximate composition. It is in fact an indefinite mixture of several or perhaps of many compound bodies, whose precise nature is little known. These bodies have received the collective names *Humus* and *Geine.* We shall employ the term *humus* to designate this mixture, whether occurring in peat, swamp-muck, salt-mud, in composts, or in the arable soil. Its chemical characters are much the same, whatever its appearance or mode of occurrence; and this is to be expected since it is always formed from the same materials and under essentially similar conditions.

Resinous and *Bituminous matters.*—If dry pulverized peat be agitated and warmed for a short time with alcohol, there is usually extracted a small amount of *resinous* and sometimes of *bituminous* matters, which are of no account in the agricultural applications of peat, but have a bearing on its value as fuel.

Ulmic and *Humic acids.* — On boiling what remains from the treatment with alcohol, with a weak solution of carbonate of soda (sal-soda), we obtain a yellowish-brown or black liquid. This liquid contains certain acid ingredients of the peat which become soluble by entering into chemical combination with soda.

On adding to the solution strong vinegar, or any other strong acid, there separates a bulky brown or black substance, which, after a time, subsides to the bottom of the vessel as a precipitate, to use a chemical term, leaving the liquid of a more or less yellow tinge. This deposit, if obtained from light brown peat, is *ulmic acid;* if from black peat, it is ***humic acid.*** These acids, when in the precipitated state, are insoluble in vinegar; but when this is washed away, they are considerably soluble in water. They are, in fact, modified by the action of the soda, so as to acquire much greater solubility in water than they otherwise possess. On drying the bulky bodies thus obtained, brown or black lustrous masses result, which have much the appearance of coal.

Ulmin and ***Humin.*** — After extracting the peat with solution of carbonate of soda, it still contains ulmin or humin. These bodies cannot be obtained in the pure state from peat, since they are mixed with more or less partially decomposed vegetable matters from which they cannot be separated without suffering chemical change. They have been procured, however, by the action of muriatic acid on sugar. They are indifferent in their chemical characters, are insoluble in water and in solution of carbonate of soda; but upon heating with solution of hydrate of soda they give dark-colored liquids, being in fact converted by this treatment into ulmic and humic acids, respectively, with which they are identical in composition.

The terms ulmic and humic acids do not refer each to a single compound, but rather to a group of bodies of closely similar appearance and properties, which, however, do differ slightly in their characteristics, and differ also in composition by containing more or less of oxygen and hydrogen in equal equivalents.

After complete extraction with hydrate of soda, there remains more or less undecomposed vegetable matter, together with sand and soil, were these contained in the peat.

Crenic and *apocrenic acids.*—From the usually yellowish liquid out of which the ulmic and humic acids have been separated, may further be procured by appropriate chemical means, not needful to be detailed here, two other bodies which bear the names respectively of *Crenic Acid* and *Apocrenic Acid.* These acids were discovered by Berzelius, the great Swedish chemist, in the water and sediment of the Porla spring, in Sweden.

By the action upon peat of carbonate of ammonia, which is generated to some extent in the decay of vegetable matters and is also absorbed from the air, ulmic and humic acids are made soluble, and combine with the ammonia as well as with lime, oxide of iron, etc. In some cases the ulmates and humates thus produced may be extracted from the peat by water, and consequently occur dissolved in the water of the swamp from which the peat is taken, giving it a yellow or brown color.

Ulmates and *Humates.*—Of considerable interest to us here, are the properties of the compounds of these acids, that may be formed in peat when it is used as an ingredient of composts. The ulmates and humates of the alkalies, viz.: *potash*, *soda*, and *ammonia*, dissolve readily in water. They are formed when the alkalies or their carbonates act on ulmin and humin, or upon ulmates or humates of lime, iron, etc. Their dilute solutions are yellow, or brown.

The ulmates and humates of *lime*, *magnesia*, oxide of *iron*, oxide of *manganese* and *alumina*, are insoluble, or nearly so in water.

In ordinary soils, the earths and oxides just named, predominate over the alkalies, and although they may contain considerable ulmic and humic acids, water is able to extract but very minute quantities of the latter, on account of the insolubility of the compounds they have formed.

On the other hand, peat, highly manured garden soil, leaf-mold, rotted manure and composts, yield yellow or brown extracts with water, from the fact that alkalies are here present to form soluble compounds.

An important fact established by Mulder is, that when solutions of alkali-carbonates are put in contact with the insoluble ulmates and humates, the latter are decomposed; soluble alkali-ulmates and humates being formed, and *in these, a portion of the otherwise insoluble ulmates and humates dissolve*, so that thus, in a compost, lime, magnesia, oxide of iron, and even alumina may exist in soluble combinations, by the agency of these acids.

Crenates and *Apocrenates.*—The ulmic and humic acids when separated from their compounds, are nearly insoluble, and, so far as we know, comparatively inert bodies; by further change, (uniting with oxygen) they pass into or yield the crenic and apocrenic acids which, according to Mulder, have an acid taste, being freely soluble in water, and in all respects, decided acids. The compounds of both these acids with the alkalies are soluble. The crenates of lime, magnesia, and protoxide of iron are soluble, crenates of peroxide of iron and of oxide of manganese are but very slightly soluble; crenate of alumina is insoluble. The apocrenates of iron and manganese are slightly soluble; those of lime, magnesia, and alumina are insoluble. All the insoluble crenates and apocrenates, are soluble in solutions of the corresponding salts of the alkalies.

Application of these facts will be given in subsequent paragraphs. It may be here remarked, that the crenate of protoxide of iron is not unfrequently formed in considerable quantity in peat-bogs, and dissolving in the water of springs gives them a chalybeate character. Copious springs of this kind occur at the edge of a peat-bed at Woodstock, Conn., which are in no small repute for their medicinal qualities, having a tonic effect from the iron they contain. Such waters, on exposure to the air, shortly absorb oxygen, and the substance is thereby converted into crenate and afterwards into apocrenate of peroxide of iron, which, being but slightly soluble, or insoluble, separates as a yellow or brown ochreous deposit along the course of the water. By further exposure to air the organic acid is oxidized to carbonic acid, and hydrated oxide of iron remains. Bog-iron ore appears often to have originated in this way.

Gein and Geic acid.—Mulder formerly believed another substance to exist in peat which he called *Gein*, and from this by the action of alkalies he supposed geic acid to be formed. In his later writings, however, he expresses doubt as to the existence of such a substance, and we may omit further notice of it, especially since, if it really do occur, its properties are not distinct from those of humic acid.

We should not neglect to remark, however, that the word gein has been employed by some writers in the sense in which we use humus, viz. : to denote the brown or black products of the decomposition of vegetable matters.

It is scarcely to be doubted that other organic compounds exist in peat. As yet, however, we have no knowledge of any other ingredients, while it appears certain that those we have described are its chief constituents, and give it its peculiar properties. With regard to them it must nevertheless be admitted, that our chemical knowledge is not entirely satisfactory, and new investigations are urgently demanded to supply the deficiencies of the researches so ably made by Mulder, more than twenty years ago.

Elementary Composition of Peat.

After this brief notice of those organic *compounds* that have been recognized in or produced from peat, we may give attention to the elementary composition of peat itself.

Like that of the vegetation from which it originates, the organic part of peat consists of Carbon, Hydrogen, Oxygen and Nitrogen. In the subjoined table are given the proportions of these elements as found in the combustible part of sphagnum, of several kinds of wood, and in that of a number of peats in various stages of ripeness. They are arranged in the order of their content of carbon.

	Analyst.	*Carbon.*	*Hydrogen.*	*Oxygen.*	*Nitrogen.*
1—Sphagnum, undecomposed.	Websky	49.88	6.54	42.42	1.16
2—Peach wood, undecomposed.	Chevandier	49.90	6.10	43.10	0.90
3—Poplar " undecomposed.	"	50.30	6.30	42.40	1.00
4—Oak " undecomposed.	"	50.60	6.00	42.10	1.30
5—Peat, porous, light-brown, sphagnous	Websky	50.86	5.80	42.57	0.77
6— " porous, red-brown.	Jaeckel	53.51	5.90	40.59	
7— " heavy, brown.	"	56.43	5.32	38.25	
8— " dark red-brown, well decomposed	Websky	59.47	6.52	31.51	2.51
9— " black, very dense and hard.	"	59.70	5.70	33.04	1.56
10— " black, heavy, best quality for fuel.	"	59.71	5.27	32.07	2.59
11— " brown, heavy, best quality for fuel.	"	62.54	6.81	29.24	1.41

From this table it is seen that sphagnum, and the wood of our forest trees are very similar in composition, though not identical. Further, it is seen from analyses 1 and 5, that in the first stages of the conversion of sphagnum into peat—which are marked by a change of color, but in which the form of the sphagnum is to a considerable extent preserved—but little alteration occurs in ultimate composition; about one *per cent.* of carbon being gained, and one of hydrogen lost. We notice in running down the columns that as the peat becomes heavier and darker in color, it also becomes richer in carbon and poorer in oxygen. Hydrogen varies but slightly.

As a general statement we may say that the ripest and heaviest peat contains 10 or 12 *per cent.* more carbon and 10 or 12 *per cent.* less oxygen than the vegetable matter from which it is produced; while between the unaltered vegetation and the last stage of humification, the peat runs through an indefinite number of intermediate stages.

Nitrogen is variable, but, in general, the older peats contain the most. To this topic we shall shortly recur, and now pass on to notice—

The ultimate composition of the compounds of which peat consists.

Below are tabulated analyses of the organic acids of peat:—

	Carbon.	*Hydrogen.*	*Oxygen.*
Ulmic acid, artificial from sugar	67.10	4.20	28.70
Humic acid, from Frisian peat	61.10	4.30	34.60
Crenic acid	56.47	2.74	40.78
Apocrenic acid	45.70	4.80	49.50

It is seen that the amount of carbon diminishes from ulmic acid to apocrenic, that of oxygen increases in the same direction and to the same extent, viz.: about 21 *per cent.*, while the hydrogen remains nearly the same in all.

(*b*) *The mineral part of peat, which remains as ashes* when the organic matters are burned away, is variable in quantity and composition. Usually a portion of sand or soil is found in it, and this not unfrequently constitutes its larger portion. Some peats leave on burning much carbonate of lime; others chiefly sulphate of lime; the ash of others again is mostly oxyd of iron; silicic, and phosphoric acids, magnesia, potash, soda, alumina and chlorine, also occur in small quantities in the ash of all peats.

With one exception (alumina) all these bodies are important ingredients of agricultural plants.

In some rare instances, peats are found, which are so impregnated with soluble sulphates of iron and alumina, as to yield these salts to water in large quantity; and sulphate of iron (green vitriol,) has actually been manufactured from such peats, which in consequence have been characterized as *vitriol peats.*

Those bases (lime, oxide of iron, etc.,) which are found as carbonates or simple oxides in the ashes, exist in the peat itself in combination with the humic and other organic acids. When these compounds are destroyed by burning, the bases remain united to carbonic acid.

5.—*Chemical Changes that occur in the formation of Peat.* When a plant perishes, its conversion into humus usually begins at once. When exposed to the atmosphere, the oxygen of the air attacks it, uniting with its carbon producing carbonic acid gas, and with its hydrogen generating water. This action goes on, though slowly, even at some depth under water, because the latter dissolves oxygen from the air in small quantity,* and constantly resupplies itself as rapidly as the gas is consumed.

Whether exposed to the air or not, the organic matter suffers internal decomposition, and portions of its elements assume the gaseous or liquid form. We have seen that ripe peat is 10 to 12 *per cent.* richer in carbon and equally poorer in oxygen, than the vegetable matters from which it originates. Organic matters, in passing into peat, lose carbon and nitrogen; but they lose oxygen more rapidly than the other two elements, and hence the latter become relatively more abundant. The loss of hydrogen is such that its proportion to the other elements is but little altered.

The bodies that separate from the decomposing vegetable matter are carbonic acid gas, carburetted hydrogen (marsh gas), nitrogen gas, and water.

Carbonic acid is the most abundant gaseous product of the peaty decomposition. Since it contains nearly 73 *per cent.* of oxygen and but 27 *per cent.* of carbon, it is ob-

* The oxygen thus absorbed by water, serves for the respiration of fish and aquatic animals.

vious that by its escape the proportion of carbon in the residual mass is increased. In the formation of water from the decaying matters, 1 part of hydrogen carries off 8 parts of oxygen, and this change increases the proportion of carbon and of hydrogen. Marsh gas consists of one part of hydrogen to three of carbon, but it is evolved in comparatively small quantity, and hence has no effect in diminishing the *per cent.* of carbon.

The gas that bubbles up through the water of a peat-bog, especially if the decomposing matters at the bottom be stirred, consists largely of marsh gas and nitrogen, often with but a small proportion of carbonic acid. Thus Websky found in gas from a peat-bed

Carbonic acid	2.97
Marsh gas	43.36
Nitrogen	53.67
	100.00

Carbonic acid, however, dissolves to a considerable extent in water, and is furthermore absorbed by the living vegetation, which is not true of marsh gas and nitrogen; hence the latter escape while the former does not. Nitrogen escapes in the uncombined state, as it always (or usually) does in the decay of vegetable and animal matters that contain it. Its loss is, in general, slower than that of the other elements, and it sometimes accumulates in the peat in considerable quantity. A small portion of nitrogen unites with hydrogen, forming ammonia, which remains combined with the humic and other acids.

PART II.

ON THE AGRICULTURAL USES OF PEAT AND SWAMP MUCK.

After the foregoing account of the composition of peat, we may proceed to notice:

1.—*The characters that adapt it for agricultural uses.*

These characters are conveniently discussed under two heads, viz.:

Those which render it useful in improving the texture and physical characters of the soil, and indirectly contribute to the nourishment of crops,—characters which constitute it an *amendment* to the soil (*A*); and

Those which make it a direct *fertilizer* (*B*).

A.—Considered as an amendment, the value of peat depends upon

Its remarkable power of absorbing and retaining water, both as a liquid and as a vapor (I):

Its power of absorbing ammonia (II):

Its effect in promoting the disintegration and solution of mineral ingredients, that is the stony matters of the soil (III)*: and*

Its influence on the temperature of the soil (IV).

The agricultural importance of these properties of peat is best illustrated by considering the faults of a certain class of soils.

Throughout the State of Connecticut, for instance, are found abundant examples of light, leachy, hungry soils, which consist of coarse sand or fine gravel; are surface-dry in a few hours after the heaviest rains, and in the summer drouths, are as dry as an ash-heap to a depth of several or many feet.

These soils are easy to work, are ready for the plow early in the spring, and if well manured give fair crops in wet seasons. In a dry summer, however, they yield poorly, or fail of crops entirely; and, at the best, they require constant and very heavy manuring to keep them in heart.

Crops fail on these soils from two causes, viz.; *want of moisture* and *want of food.* Cultivated plants demand as an indispensable condition of their growth and perfection, to be supplied with water in certain quantities, which differ with different crops. Buckwheat will flourish best on dry soils, while cranberries and rice grow in swamps.

Our ordinary cereal, root, forage and garden crops require a medium degree of moisture, and with us it is in all cases desirable that the soil be equally protected from excess of water and from drouth. Soils must be thus situated either naturally, or as the result of improvement, before any steadily good results can be obtained in their cultivation. The remedy for excess of water in too heavy soils, is thorough drainage. It is expensive, but effectual. It makes the earth more porous, opens and maintains channels, through which the surplus water speedily runs off, and permits the roots of crops to go down to a considerable depth.

What, let us consider, is the means of obviating the defects of soils that are naturally too porous, from which the water runs off too readily, and whose crops "burn up" in dry seasons?

In wet summers, these light soils, as we have remarked, are quite productive if well manured. It is then plain that if we could add anything to them which would retain the moisture of dews and rains in spite of the summer-heats, our crops would be uniformly fair, provided the supply of manure were kept up.

But why is it that light soils, need more manure than loamy or heavy lands? We answer—because, in the first place, the rains which quickly descend through the open soil, wash down out of the reach of vegetation the soluble fertilizing matters, especially the nitrates, for which the soil has no retentive power; and in the second place, from the porosity of the soil, the air has too great access, so that the vegetable and animal matters of manures decay too rapidly, their volatile portions, ammonia and carbonic acid, escape into the atmosphere, and are in measure lost to the crops. From these combined causes we find that a heavy dressing of well-rotted stable manure, almost if not entirely, disappears from such soils in one season, so that another year the field requires a renewed application; while on loamy soils the same amount of manure would have lasted several years, and produced each year a better effect.

We want then to *amend* light soils by incorporating with them something that prevents the rains from leaching through them too rapidly, and also that renders them less open to the air, or absorbs and retains for the use of crops the volatile products of the decay of manures.

For these purposes, vegetable matter of some sort is the best and almost the only amendment that can be economically employed. In many cases a good peat or muck is the best form of this material, that lies at the farmer's command.

I.—*Its absorbent power for liquid water* is well known to every farmer who has thrown it up in a pile to season for use. It holds the water like a sponge, and, according to its greater or less porosity, will retain from 50 to 100 or more *per cent.* of its weight of liquid, without dripping. Nor can this water escape from it rapidly. It dries almost as slowly as clay, and a heap of it that has been exposed to sun and wind for a whole summer, though it has of course lost much water, is still distinctly wet to the eye and the feel a little below the surface.

Its absorbent power for vapor of water is so great that more than once it has happened in Germany, that barns or close sheds filled with partially dried peat, such as is used for fuel, have been burst by the swelling of the peat in damp weather, occasioned by the absorption of moisture from the air. This power is further shown by the fact that when peat has been kept all summer long in a warm room, thinly spread out to the air, and has become like dry snuff to the feel, it still contains from 8 to 30 *per cent.* (average 15 *per cent.*) of water. To dry a peat thoroughly, it requires to be exposed for some time to the temperature of boiling water. It is thus plain, as experience has repeatedly demonstrated, that no ordinary summer heats can dry up a soil which has had a good dressing of this material, for on the one hand, it soaks up and holds the rains that fall upon it, and on the other, it absorbs the vapor of water out of the atmosphere whenever it is moist, as at night and in cloudy weather.

When peat has once become *air-dry*, it no longer manifests this avidity for water. In drying it shrinks, loses its porosity and requires long soaking to saturate it again. In the soil, however, it rarely becomes air-dry, unless indeed, this may happen during long drouth with a peaty soil, such as results from the draining of a bog.

II.—*Absorbent power for ammonia.*

All soils that deserve to be called fertile, have the property of absorbing and retaining ammonia and the volatile matters which escape from fermenting manures, but light and coarse soils may be deficient in this power. Here again in respect to its absorptive power for ammonia, peat comes to our aid.

It is easy to show by direct experiment that peat absorbs and combines with ammonia.

In 1858 I took a weighed quantity of air-dry peat from the New Haven Beaver Pond, (a specimen furnished me by Chauncey Goodyear, Esq.,) and poured upon it a known quantity of dilute solution of ammonia, and agitated the two together occasionally during 48 hours. I then distilled off at a boiling heat the unabsorbed ammonia and determined its quantity. This amount subtracted from that of the ammonia originally employed, gave the quantity of ammonia absorbed and retained by the peat at the temperature of boiling water.

The peat retained ammonia to the amount of 0.95 of one *per cent.*

I made another trial at the same time with carbonate of ammonia, adding excess of solution of this salt to a quantity of peat, and exposing it to the heat of boiling water, until no smell of ammonia was perceptible. The entire nitrogen in the peat was then determined, and it was found that the dry peat which originally contained nitrogen equivalent to 2.4 *per cent.* of ammonia, now yielded an amount corresponding to 3.7 *per cent.* The quantity of ammonia absorbed and retained at a temperature of 212°, was thus 1.3 *per cent.*

This last experiment most nearly represents the true power of absorption; because, in fermenting manures, ammonia mostly occurs in the form of carbonate, and this is more largely retained than free ammonia, on account of its power of decomposing the humate of lime, forming with it carbonate of lime and humate of ammonia.

The absorbent power of peat is well shown by the analyses of three specimens, sent me in 1858, by Edwin Hoyt, Esq., of New Canaan, Conn. The first of these was the swamp muck he employed. It contained in the air-dry state nitrogen equivalent to 0.58 *per cent.* of ammonia. The second sample was the same muck that had lain under the flooring of the horse stables, and had been, in this way, partially saturated with urine. It contained nitrogen equivalent to 1.15 *per cent.* of ammonia. The third sample was, finally, the same muck composted with white-fish. It contained nitrogen corresponding to 1.31 *per cent.* of ammonia.*

The quantities of ammonia thus absorbed, both in the laboratory and field experiments are small — from 0.7 to 1.3 *per cent.* The absorption is without doubt chiefly due to the organic matter of the peats, and in all the specimens on which these trials were made, the proportion of inorganic matter is large. The results therefore become a better expression of the power of *peat*, in general, to absorb ammonia, if we reckon them on the organic matter alone. Calculated in this way, the organic matter of the Beaver Pond peat (which constitutes but 68 *per cent.* of the dry peat) absorbs 1.4 *per cent.* of free ammonia, and 1.9 *per cent.* of ammonia out of the carbonate of ammonia.

* This sample contained also fish-bones, hence the larger content of nitrogen was not entirely due to absorbed ammonia.

Similar experiments, by Anderson, on a Scotch peat, showed it to possess, when wet, an absorptive power of 2 *per cent.*, and, after drying in the air, it still retained 1.5 *per cent.*—[Trans. Highland and Ag'l Soc'y.]

When we consider how small an ingredient of most manures nitrogen is, viz.: from one-half to three-quarters of one *per cent.* in case of stable manure, and how little of it, in the shape of guano for instance, is usually applied to crops—not more than 40 to 60 lbs. to the acre, (the usual dressings with guano are from 250 to 400 lbs. per acre, and nitrogen averages but 15 *per cent.* of the guano), we at once perceive that an absorptive power of one or even one-half *per cent.* is greatly more than adequate for every agricultural purpose.

III.—*Peat promotes the disintegration of the soil.*

The soil is a storehouse of food for crops; the stores it contains are, however, only partly available for immediate use. In fact, by far the larger share is locked up, as it were, in insoluble combinations, and only by a slow and gradual change can it become accessible to the plant. This change is largely brought about by the united action of *water* and *carbonic acid gas.* Nearly all the rocks and minerals out of which fertile soils are formed,—which therefore contain those inorganic matters that are essential to vegetable growth,—though very slowly acted on by pure water, are decomposed and dissolved to a much greater extent by water, charged with carbonic acid gas.

It is by these solvents that the formation of soil from broken rocks is to a great extent due. Clay is invariably a result of their direct action upon rocks. The efficiency of the soil depends greatly upon their chemical influence.

The only abundant source of carbonic acid in the soil, is decaying vegetable matter.

Hungry, leachy soils, from their deficiency of vegetable matter and of moisture, do not adequately yield their own native resources to the support of crops, because the conditions for converting their fixed into floating capital are wanting. Such soils dressed with peat or green manured, at once acquire the power of retaining water, and keep that water ever charged with carbonic acid: thus not only the extraneous manures which the farmer applies are fully economized; but the soil becomes more productive from its own stores of fertility which now begin to be unlocked and available.

Dr. Peters, of Saxony, has made some instructive experiments that are here in point. He filled several large glass jars, (2½ feet high and 5½ inches wide) with a rather poor loamy sand, containing considerable humus, and planted in each one, June 14, 1857, an equal number of seeds of oats and peas. Jar No. 2 had daily passed into it through a tube, adapted to the bottom, about 3¼ pints of common air. No. 3 received daily the same bulk of a mixture of air and carbonic acid gas, of which the latter amounted to one-fourth. No. 1 remained without any treatment of this kind, *i. e.*: in just the condition of the soil in an open field, having no air in its pores, save that penetrating it from the atmosphere. On October 3, the plants were removed from the soil, and after drying at the boiling point of water, were weighed. The crops from the pots into which air and carbonic acid were daily forced, were about *twice as heavy* as No. 1, which remained in the ordinary condition.

Examination of the soil further demonstrated, that in the last two soils, a considerably greater quantity of mineral and organic matters had become soluble in water, than in the soil that was not artificially aërated. The actual results are given in the table below in grammes, and refer to 6000 grammes of soil in each case: —

ACTION OF CARBONIC ACID ON THE SOIL.

Substances soluble in water, etc.	*No. 1, Without Artificial Supply of Air.*	*No. 2, Common Air Added.*	*No. 3, Air and Carbonic acid added*
Mineral matters	2.04	3.71	4.99
Potash	0.07	0.17	0.14
Soda	0.17	0.23	0.28
Organic matters	2.76	4.32	2.43
Weight of Crops	5.89	10.49	12.35

It will be seen from the above that air alone exercised nearly as much solvent effect as the mixture of air with one-fourth its weight of carbonic acid; this is doubtless, in part due to the fact that the air, upon entering the soil rich in humus, caused the abundant formation of carbonic acid, as will be presently shown must have been the case. It is, however, probable that organic acids (crenic and apocrenic,) and nitric acid were also produced (by oxidation,) and shared with carbonic the work of solution.

It is almost certain, that the acids of peat exert a powerful decomposing, and ultimately solvent effect on the minerals of the soil; but on this point we have no precise information, and must therefore be content merely to present the probability. This is sustained by the fact that the crenic, apocrenic and humic acids, though often partly uncombined, are never wholly so, but usually occur united in part to various bases, viz.: lime, magnesia, ammonia, potash, alumina and oxide of iron.

The crenic and apocrenic acids (that are formed by the oxidation of ulmic and humic acids,) have such decided acid characters, — crenic acid especially, which has a strongly sour taste—that we cannot well doubt their dissolving action.

IV.—*The influence of peat on the temperature* of light soils dressed with it may often be of considerable practical importance. A light dry soil is subject to great variations of temperature, and rapidly follows the changes of the atmosphere from cold to hot, and from hot to cold. In the summer noon a sandy soil becomes so warm as to be hardly endurable to the feel, and again it is on such soils that the earliest frosts take effect. If a soil thus subject to extremes of temperature have a dressing of peat, it will on the one hand not become so warm in the hot day, and on the other hand it will not cool so rapidly, nor so much in the night; its temperature will be rendered more uniform, and on the whole, more conducive to the welfare of vegetation. This regulative effect on temperature is partly due to the stores of water held by peat. In a hot day this water is constantly evaporating, and this, as all know, is a cooling process. At night the peat absorbs vapor of water from the air, and condenses it within its pores, this condensation is again accompanied with the evolution of heat.

It appears to be a general, though not invariable fact, that dark colored soils, other things being equal, are constantly the warmest, or at any rate maintain the temperature most favorable to vegetation. It has been repeatedly observed that on light-colored soils plants mature more rapidly, if the earth be thinly covered with a coating of some black substance. Thus Lampadius, Professor in the School of Mines at Freiberg, a town situated in a mountainous part of Saxony, found that he could ripen melons, even in the coolest summers, by strewing a coating of coal-dust an inch deep over the surface of the soil. In some of the vineyards of the Rhine, the powder of a black slate is employed to hasten the ripening of the grape.

Girardin, an eminent French agriculturist, in a series of experiments on the cultivation of potatoes, found that the time of their ripening varied eight to fourteen days, according to the character of the soil. He found, on the 25th of August, in a very dark soil, made so by the presence of much humus or decaying vegetable matter, twenty-six varieties ripe; in sandy soil but twenty, in clay nineteen, and in a white lime soil only sixteen.

It cannot be doubted then, that the effect of dressing a light sandy or gravelly soil with peat, or otherwise enriching it in vegetable matter, is to render it warmer, in the sense in which that word is usually applied to soils. The upward range of the thermometer is not, indeed, increased, but the uniform warmth so salutary to our most valued crops is thereby secured.

In the light soils stable-manure wastes too rapidly because, for one reason, at the extremes of high temperature, oxidation and decay proceed with great rapidity, and the volatile portions of the fertilizer are used up faster than the plant can appropriate them, so that not only are they wasted during the early periods of growth, but they are wanting at a later period when their absence may prove the failure of a crop.

B. The ingredients and qualities which make peat *a direct fertilizer* next come under discussion. We shall notice:

The organic matters, including nitrogen (*ammonia and nitric acid*) (I):

The inorganic or mineral ingredients (II):

Peculiarities in the decay of Peat (III), *and*

Institute a comparison between peat and stable manure (IV).

I.—Under this division we have to consider:

1. *The organic matters as direct food to plants.*

Thirty years ago, when Chemistry and Vegetable Physiology began to be applied to Agriculture, the opinion was firmly held among scientific men, that the organic parts of humus—by which we understand decayed vegetable matter, such as is found to a greater or less extent in all good soils, and *abounds* in many fertile ones, such as constitutes the leaf-mold of forests, such as is produced in the fermenting of stable manure, and that forms the principal part of swamp-muck and peat,—are the true nourishment of vegetation, at any rate of the higher orders of plants, those which supply food to man and to domestic animals.

In 1840, Liebig, in his celebrated treatise on the "Applications of Chemistry to Agriculture and Physiology," gave as his opinion that these organic bodies do not nourish vegetation except by the products of their decay. He asserted that they cannot enter the plant directly, but that the water, carbonic acid and ammonia resulting from their decay, are the substances actually imbibed by plants, and from these alone is built up the organic or combustible part of vegetation.

To this day there is a division of opinion among scientific men on this subject, some adopting the views of Liebig, others maintaining that certain soluble organic matters, viz., crenic and apocrenic acids are proper food of plants.

On the one hand it has been abundantly demonstrated that these organic matters are not at all essential to the growth of agricultural plants, and can constitute but a small part of the actual food of vegetation taken in the aggregate.

On the other hand, we are acquainted with no satisfactory evidence that the soluble organic matters of the soil and of peat, especially the crenates and apocrenates, are not actually appropriated by, and, so far as they go, are not directly serviceable as food to plants.

Be this as it may, practice has abundantly demonstrated the value of humus as an ingredient of the soil, and if not directly, yet indirectly, it furnishes the material out of which plants build up their parts.

2. *The organic matters of peat as indirect food to plants.* Very nearly one-half, by weight, of our common crops, when perfectly dry, consists of *carbon.* The substance which supplies this element to plants is the gas, carbonic acid. Plants derive this gas mostly from the atmosphere, absorbing it by means of their leaves. But the free atmosphere, at only a little space above the soil, contains on the average but $\frac{1}{2500}$ of its bulk of this gas, whereas plants flourish in air containing a larger quantity, and, in fact, their other wants being supplied, they grow better as the quantity is increased to $\frac{1}{12}$ the bulk of the air. These considerations make sufficiently obvious how important it is that the soil have in itself a constant and abundant source of carbonic acid gas. As before said, *organic matter, in a state of decay,* is the single material which the farmer can incorporate with his soil in order to make the latter a supply of this most indispensable form of plant-food.

When organic matters decay in the soil, their carbon ultimately assumes the form of Carbonic acid. This gas, constantly exhaling from the soil, is taken up by the foliage of the crops, and to some extent is absorbed likewise by their roots.

Boussingault & Lewy have examined the air inclosed in the interstices of various soils, and invariably found it much richer (10 to 400 times) than that of the atmosphere above. Here follow some of their results:

CARBONIC ACID IN SOILS.

Designation and Condition of Soil.	*Volumes of Carbonic acid in 100 of air in pores of Soil.*	*Cubic feet of air in acre to depth of 14 inches.*	*Cubic feet of Carbonic acid in acre to depth of 14 inches.*
Sandy subsoil of forest	0.24	4,326	14
Loamy " " "	0.82	3,4[illegible]8	[illegible]8
Surface soil " "	0.86	5,768	56
Clayey soil of artichoke field	0.66	10,094	71
Soil of asparagus bed, unmanured for one year	0.79	10,948	86
" " " " newly manured	1.54	10,948	172
Sandy soil, six days after manuring, and three days of rain	2.21	11,536	257
" " ten " " " " " " "	9.74	11.536	1144
Compost of vegetable mold	3.64	20,608	772
Carbonic Acid in Atmosphere.	*Volumes of Carbonic acid to 100 of air above the soil.*	*Cubic feet of air over one acre to height of 14 inches.*	*Cubic feet of Carbonic acid over one acre to height of 14 inches.*
	0.025	50,820	14

From the above it is seen that in soils containing little decomposing organic matters—as the forest sub-soils—the quantity of carbonic acid is no greater than that contained in an equal bulk of the atmosphere. It is greater in loamy and clayey soils; but is still small. In the artichoke field (probably light soil not lately manured), and even in an asparagus bed unmanured for one year, the amount of carbonic acid is not greatly larger. In newly manured fields, and especially in a vegetable compost, the quantity is vastly greater.

The organic matters which come from manures, or from the roots and other residues of crops, are the source of the carbonic acid of the soil. These matters continually waste in yielding this gas, and must be supplied anew. Boussingault found that the rich soil of his kitchen-garden (near Strasburg) which had been heavily manured from the barn-yard for many years, lost one-third of its carbon by exposure to the air for three months (July, August and September,) being daily watered. It originally contained 2.43 *per cent.* At the conclusion of the experiment it contained but 1.60 *per cent*, having lost 0.83 *per cent.*

Peat and swamp-muck, when properly prepared, furnish carbonic acid in large quantities during their slow oxidation in the soil.

3. *The Nitrogen of Peat, including Ammonia and Nitric Acid.*

The sources of the nitrogen of plants, and the real cause of the value of nitrogenous fertilizers, are topics that have excited more discussion than any other points in Agricultural Chemistry. This is the result of two circumstances. One is the obscurity in which some parts of the subject have rested; the other is the immense practical and commercial importance of this element, as a characteristic and essential ingredient of the most precious fertilizers. It is a rule that the most valuable manures, *commercially considered*, are those containing the most nitrogen. Peruvian guano, sulphate of ammonia, soda-saltpeter, fish and flesh manures, bones and urine, cost the farmer more money per ton than any other manures he buys or makes, superphosphate of lime excepted, and this does not find sale, for general purposes, unless it contains several *per cent.* of nitrogen. These are, in the highest sense, nitrogenous fertilizers, and, if deprived of their nitrogen, they would lose the greater share of their fertilizing power.

The importance of the nitrogen of manures depends upon the fact that those forms (compounds) of nitrogen which are capable of supplying it to vegetation are comparatively scarce.

It has long been known that peat contains a considerable quantity of nitrogen. The average amount in thirty specimens, analyzed under the author's direction, including peats and swamp mucks of all grades of quality, is equivalent to 1½ *per cent.* of the air-dried substance, or more than thrice as much as exists in ordinary stable or yard manure. In several peats the amount is as high as 2.4 *per cent.*, and in one case 2.9 *per cent.* were found.

Of these thirty samples, one-half were largely mixed with soil, and contained from 15 to 60 *per cent.* of mineral matters.

Reducing them to an average of 15 *per cent.* of water and 5 *per cent.* of ash, they contain 2.1 *per cent.* of nitrogen, while the organic part, considered free from water and mineral substances, contains on the average 2.6 *per cent.* See table, page 90.

The five peats, analyzed by Websky and Chevandier, as cited on page 24, considered free from water and ash, contain an average of 1.8 *per cent.* of nitrogen.

We should not neglect to notice that peat is often comparatively poor in nitrogen. Of the specimens, examined in the Yale Analytical Laboratory, several contained but half a *per cent.* or less. So in the analyses of Websky, one sample contained but 0.77 *per cent.* of the element in question.

As concerns the state of combination in which nitrogen exists in peat, there is a difference of opinion. Mulder regards it as chiefly occurring in the form of *ammonia* (a compound of nitrogen and hydrogen), united to the organic acids from which it is very difficult to separate it. Recent investigations indicate that in general, peat contains but a small proportion of ready-formed ammonia.

The great part of the nitrogen of peat exists in an insoluble and inert form: but, by the action of the atmosphere upon it, especially when mixed with and divided by the soil, it gradually becomes available to vegetation to as great an extent as the nitrogen of ordinary fertilizers.

It appears from late examinations that weathered peat may contain *nitric acid* (compound of nitrogen with oxygen) in a proportion which, though small, is yet of great importance, agriculturally speaking. What analytical data we possess are subjoined.

PROPORTIONS OF NITROGEN, ETC., IN PEAT.

		Analyst.	*Total Nitrogen.*	*Ammonia, per cent.*	*Nitric acid.*
1—Brown Peat......	Air dry (?)	Boussingault	2.20	0.018	0.000
2—Black Peat	"	"	Undetermined	0.025	Undetermined
3—Peat	Dried at 212°	Reichardt*	"	0.152	0.483
4—Peat	"	"	"	0.165	0.525
5—Peat	"	"	"	0.305	0.241
6—Peat	"	"	"	0.335	0.421

Specimens 3, 4 and 5, are swamp (or heath) mucks, and have been weathered for use in flower-culture. 3 and 4 are alike, save that 3 has been weathered a year longer than 4. They contain respectively 41, 56 and 67 *per cent.* of organic matter.

Sample 6, containing 86 *per cent.* of organic matter, is employed as a manure with great advantage, and probably was weathered before analysis. It contained 85 *per cent.* of organic substance.

More important to us than the circumstance that this peat contains but little or no ammonia or nitric acid, and the other contains such or such a fraction of one *per cent.* of these bodies, is the grand fact that all peats may yield a good share of their nitrogen to the support of crops, when properly treated and applied.

Under the influence of Liebig's teachings, which were logically based upon the best data at the disposal of this distinguished philosopher when he wrote 25 years ago, it

* Reichardt's analyses are probably inaccurate, and give too much ammonia and nitric acid.

has been believed that the nitrogen of a fertilizer, in order to be available, must be converted into ammonia and presented in that shape to the plant. It has been recently made clear that nitric acid, rather than ammonia, is the form of nitrogenous food which is most serviceable to vegetation, and the one which is most abundantly supplied by the air and soil. The value of ammonia is however positive, and not to be overlooked.

When peat, properly prepared by weathering or composting, is suitably incorporated with a poor or light soil, it slowly suffers decomposition and wastes away. If it be wet, and air have access in limited quantity, especially if *lime* be mixed with it, a portion of its nitrogen is gradually converted into ammonia. With full access of air *nitric acid* is produced. In either case, it appears that a considerable share of the nitrogen escapes in the free state as gas, thereby becoming useless to vegetation until it shall have become converted again into ammonia or nitric acid. It happens in a cultivated soil that the oxygen of the air is in excess at the surface, and less abundant as we go down until we get below organic matters: it happens that one day it is saturated with water more or less, and another day it is dry, so that at one time we have the conditions for the formation of ammonia, and at another, those favorable to producing nitric acid. In this way, so far as our present knowledge warrants us to affirm, organic matters, decaying in the soil, continuously yield portions of their nitrogen in the forms of ammonia and nitric acid for the nourishment of plants.

The farmer who skillfully employs as a fertilizer a peat containing a good proportion of nitrogen, may thus expect to get from it results similar to what would come from the corresponding quantity of nitrogen in guano or stable manure.

But the capacity of peat for feeding crops with nitrogen appears not to stop here. Under certain conditions, *the free nitrogen of the air which cannot be directly appropriated by vegetation, is oxidized in the pores of the soil to nitric acid, and thus, free of expense to the farmer, his crops are daily dressed with the most precious of all fertilizers.*

This gathering of useless nitrogen from the air, and making it over into plant-food cannot go on in a soil destitute of organic matter, requires in fact that vegetable remains or humified substances of some sort be present there. The evidence of this statement, whose truth was maintained years ago as a matter of opinion by many of the older chemists, has recently become nearly a matter of demonstration by the investigations of Boussingault and Knop, while the explanation of it is furnished by the researches of Schœnbein and Zabelin. To attempt any elucidation of it here would require more space than is at our disposal.

It is plain from the contents of this paragraph that peat or swamp muck is, in general, an abundant source of nitrogen, and is often therefore an extremely cheap means of replacing the most rare and costly fertilizers.

II.—With regard to the *inorganic matters of peat* considered as food to plants, it is obvious, that, leaving out of the account for the present, some exceptional cases, they are useful as far as they go.

In the ashes of peats, we almost always find small quantities of sulphate of lime, magnesia and phosphoric acid. Potash and soda too, are often present, though rarely to any considerable amount. Carbonate and sulphate of lime are large ingredients of the ashes of about one-half of the thirty-three peats and swamp mucks I have examined. The ashes of the other half are largely mixed with sand and soil, but in most cases also contain considerable sulphate of lime, and often carbonates of lime and magnesia.

In one swamp-muck, from Milford, Conn., there was found but two *per cent.* of ash, at least one-half of which was sand, and the remainder sulphate of lime, (gypsum.) In other samples 20, 30, 50 and even 60 *per cent.* remained after burning off the organic matter. In these cases the ash is chiefly sand. The amount of ash found in those peats which were most free from sand, ranges from five to nine *per cent.* Probably the average proportion of true ash, viz.: that derived from the organic matters themselves, not including sand and accidental ingredients, is not far from five *per cent.*

In twenty-two specimens of European peat, examined by Websky, Jæckel, Walz, Wiegmann, Einhof and Berthier, eleven contained from 0.6 to 3.5 *per cent.* of ash. The other eleven yielded from 5.3 to 22 *per cent.* The average of the former was 2.4, that of the latter 12.7 *per cent.* Most of these contained a considerable proportion of sand or soil.

Variation in the composition as well as in the quantity of ash is very great.

Three analyses of peat-ashes have been executed at the anthor's instance with the subjoined results:

ANALYSES OF PEAT-ASHES.

	A.	B.	C.
Potash	0.69	0.80	3.46
Soda	0.58		trace.
Lime	40.52	35.59	6.60
Magnesia	6.06	4.92	1.05
Oxide of iron and alumina	5.17	9.08	15.59
Phosphoric acid	0.50	0.77	1.55
Sulphuric acid	5.52	10.41	4.04
Chlorine	0.15	0.43	0.70
Soluble silica	8.23	1.40	67.01 (silica, carbonic acid and sand together)
Carbonic acid	19.60	22.28	
Sand	12.11	15.04	
	99.13	100.74	100.00

A was furnished by Mr. Daniel Buck, Jr., of Poquonnock, Conn., and comes from a peat which he uses as fuel.

B was sent by Mr. J. H. Stanwood, of Colebrook, Conn.

C was sent from Guilford, Conn., by Mr. Andrew Foote.*

A and B, after excluding sand, are seen to consist chiefly of carbonates and sulphates of lime and magnesia. III. contains a very large proportion of sand and soluble silica, much iron and alumina, less lime and sulphuric acid. Potash and phosphoric acid are three times more abundant in C than in the others.

Instead of citing in full the results of Websky, Jæckel and others, it will serve our object better to present the maximum, minimum and average proportions of the important ingredients in twenty-six recent analyses, (including these three,) that have come under the author's notice.

VARIATIONS AND AVERAGES IN COMPOSITION OF PEAT-ASHES.

	Minimum.		*Maximum.*	*Average.*	
Potash	0.05	to	3.64	0.89	per cent.
Soda	none	"	5.73	0.83	"
Lime	4 72	"	58.38	24.00	"
Magnesia	none	"	24.39	3.20	"
Alumina	0.99	"	20.50	5.78	"
Oxide of iron	none	"	73.33	18.70	"
Sulphuric acid	none	"	37.40	7.50	"
Chlorine	"	"	6.50	0.60	"
Phosphoric acid	"	"	6.29	2.56	"
Sand	0.99	"	56.97	25.50	"

It is seen from the above figures that the ash of peat varies in composition to an indefinite degree. Lime is the only ingredient that is never quite wanting, and with the exception of sand, it is on the average the largest. Of the other agriculturally valuable components, sulphuric acid has the highest average; then follows magnesia; then phosphoric acid, and lastly, potash and soda: all of these, however, may be nearly or quite lacking.

* These analyses were executed — A by Professor G. F. Barker; B by Mr. O. C. Sparrow; C by Mr. Peter Collier.

Websky, who has recently made a study of the composition of a number of German peats, believes himself warranted to conclude that peat is so modified in appearance by its mineral matters, that the quantity or character of the latter may be judged of in many cases by the eye. He remarks, (*Journal fuer Praktische Chemie*, *Bd.* 92, *S.* 87,) "that while for example the peats containing much sand and clay have a red-brown powdery appearance, and never assume a lustrous surface by pressure; those which are very rich in lime, are black, sticky when moist, hard and of a waxy luster on a pressed surface, when dry: a property which they share indeed with very dense peats that contain little ash. Peats impregnated with iron are easily recognized. Their peculiar odor, and their changed appearance distinguish them from all others."

From my own investigations on thirty specimens of Connecticut peats, I am forced to disagree with Websky entirely, and to assert that except as regards sand, which may often be detected by the eye, there is no connexion whatever between the quantity or character of the ash and the color, consistency, density or any other external quality of the peat.

The causes of this variation in the ash-content of peat, deserve a moment's notice. The plants that produce peat contain considerable proportions of lime, magnesia, alkalies, sulphuric acid, chlorine and phosphoric acid, as seen from the following analysis by Websky.

COMPOSITION OF THE ASH OF SPHAGNUM.

Potash	17.2
Soda	8.3
Lime	11.8
Magnesia	6.7
Sulphuric acid	6.5
Chlorine	6.2
Phosphoric acid	6.7

Per cent. of ash, 2.5.

The mineral matters of the sphagnum do not all become ingredients of the peat; but, as rapidly as the moss decays below, its soluble matters are to a great degree absorbed by the vegetation, which is still living and growing above. Again, when a stream flows through a peat-bed, soluble matters are carried away by the water, which is often dark-brown from the substances dissolved in it. Finally the soil of the adjacent land is washed or blown upon the swamp, in greater or less quantities.

III.—*The decomposition of peat in the soil offers some peculiarities* that are worthy of notice in this place. Peat is more gradual and regular in decay than the vegetable matters of stable dung, or than that furnished by turning under sod or green crops. It is thus a more steady and lasting benefit, especially in light soils, out of which ordinary vegetable manures disappear too rapidly. The decay of peat appears to proceed through a regular series of steps. In the soil, especially in contact with soluble alkaline bodies, as ammonia and lime, there is a progressive conversion of the *insoluble* or *less soluble* into *soluble* compounds. Thus the inert matters that resist the immediate solvent power of alkalies, absorb oxygen from the air, and form the humic or ulmic acids soluble in alkalies; the humic acids undergo conversion into crenic acid, and this body, by oxidation, passes into apocrenic acid. The two latter are soluble in water, and, in the porous soil, they are rapidly brought to the end-results of decay, viz.: water, carbonic acid, ammonia and free nitrogen.

Great differences must be observed, however, in the rapidity with which these changes take place. Doubtless they go on most slowly in case of the fibrous compact peats, and perhaps some of the lighter and more porous samples of swamp muck, would decay nearly as fast as rotted stable dung.

It might appear from the above statement, that the effect of exposing peat to the air, as is done when it is incorporated with the soil, would be to increase relatively the amount of soluble organic matters; but the truth is, that they are often actually diminished. In fact, the oxidation and consequent removal of these soluble matters (crenic and apocrenic acids,) is likely to proceed more rapidly than they can be produced from the less soluble humic acid of the peat.

IV.—*Comparison of Peat with Stable Manure.*

The fertilizing value of peat is best understood by comparing it with some standard manure. Stable manure is obviously that fertilizer whose effects are most universally observed and appreciated, and by setting analyses of the two side by side, we may see at a glance, what are the excellencies and what the deficiencies of peat. In order rightly to estimate the worth of those ingredients which occur in but small proportion in peat, we must remember that it, like stable manure, may be, and usually should be, applied in large doses, so that in fact the smallest ingredients come upon an acre in considerable quantity. In making our comparison, we will take the analysis of Peat from the farm of Mr. Daniel Buck, Jr., of Poquonnock, Conn., and the average of several analyses of rotted stable dung of *good quality.*

No. *I*, is the analysis of Peat; No. *II*, that of well rotted stable manure:—

		I.	*II.*
Water expelled at 212 degrees		79.000	79.00
Org. matter	Soluble in dilute solution of carbonate of soda	7.312	14.16
	Insoluble in solution of carbonate of soda	12.210	
Potash		0.010	0.65
Soda		0.009	
Lime		0.608	0.57
Magnesia		0.091	0.19
Phosphoric acid		0.008	0.23
Sulphuric acid		0.082	0.27
Nitrogen		0.600	0.55
Matters, soluble in water		0.450	4.42

To make the comparison as just as possible, the peat is

calculated with the same content of water, that stable dung usually has.

We observe then, that the peat contains in a given quantity, *about one-third more organic matter, an equal amount of lime and nitrogen;* but is *deficient in potash, magnesia, phosphoric and sulphuric acids.*

The deficiencies of this peat in the matter of composition may be corrected, as regards potash, by adding to 100 lbs. of it 1 lb. of potash of commerce, or 5 lbs. of unleached wood-ashes; as regards phosphoric and sulphuric acids, by adding 1 lb. of good super-phosphate, or 1 lb. each of bone dust and plaster of Paris.

In fact, the additions just named, will convert *any fresh peat*, containing not more than 80 *per cent.* of water and not less than 20 *per cent.* of organic matter, into a mixture having as much fertilizing matters as stable dung, with the possible exception of nitrogen.

It is a fact, however, that two manures may reveal to the chemist the same composition, and yet be very unlike in their fertilizing effects, because their conditions are unlike, because they differ in their degrees of solubility or availability.

As before insisted upon, it is true in general, that peat is more slow of decomposition than yard-manure, and this fact, which is an advantage in an amendment, is a disadvantage in a fertilizer. Though there may be some peats, or rather swamp mucks, which are energetic and rapid in their action, it seems that they need to be applied in larger quantities than stable manure in order to produce corresponding fertilizing effects. In many cases peat requires some preparation by weathering, or by chemical action—"fermentation"—induced by decomposing animal matters or by alkalies. This topic will shortly be discussed.

We adopt, as a general fact, the conclusion that peat is inferior in fertilizing power to stable manure.

Experience asserts, however, with regard to some individual kinds, that they are equal to common yard manure without any preparation whatever.

Mr. Daniel Buck, of Poquonnock, Conn., says, of the 'muck,' over-lying the peat, whose composition has just been compared with stable manure, that it "has been applied fresh to meadow with good results; the grass is not as tall but thicker and finer, and of a darker green in the spring, than when barn-yard manure is spread on."

A swamp muck, from Mr. A. M. Haling, Rockville, Conn., "has been used as a top-dressing, on grass, with excellent results. It is a good substitute for barn-yard manure."

A peat, from Mr. Russel U. Peck, of Berlin, Conn., "has been used fresh, on corn and meadow, with good effect."

Of the peat, from the 'Beaver Pond,' near New Haven, Mr. Chauncey Goodyear, says, "it has been largely used in a fresh state, and in this condition is as good as cow-dung."

Mr. Henry Keeler, remarks, concerning a swamp muck occurring at South Salem, N. Y., that "it has been used in the fresh state, applied to corn and potatoes, and appears to be equal to good barn manure:" further: — "it has rarely been weathered more than two months, and then applied side by side with the best yard manure has given equally good results."

A few words as to the apparent contradiction between Chemistry, which says that peat is not equal to stable dung as a fertilizer, and Practice, which in these cases affirms that it is equal to our standard manure.

In the first place, the chemical conclusion is a general one, and does not apply to individual peats, which, in a few instances, may be superior to yard manure. The practical judgment also is, that, in general, yard manure is the best.

To go to the individual cases; second: A peat in which nitrogen exists in as large a proportion as is found in stable or yard manure, being used in larger quantity, or being more durable in its action, may for a few seasons produce better results than the latter, merely on account of the presence of this one ingredient, it may in fact, for the soil and crop to which it is applied, be a better fertilizer than yard manure, because nitrogen is most needed in that soil, and yet for the generality of soils, or in the long run, it may prove to be an inferior fertilizer.

Again; third—the melioration of the physical qualities of a soil, the amendment of its dryness and excessive porosity, by means of peat, may be more effective for agricultural purposes, than the application of tenfold as much fertilizing, *i. e.* plant-feeding materials; in the same way that the mere draining of an over-moist soil often makes it more productive than the heaviest manuring.

2.—*On the characters of Peat that are detrimental, or that may sometimes need correction before it is agriculturaliy useful.*

I.—*Bad effects on wet heavy soils.*

We have laid much stress on the amending qualities of peat, when applied to dry and leachy soils, which by its use are rendered more retentive of moisture and manure. These properties, which it would seem, are just adapted to renovate very light land, under certain circumstances, may become disadvantageous on heavier soils. On clays no application is needed to retain moisture. They are already too wet as a general thing.

Peat, when put into the soil, lasts much longer than stubble, or green crops plowed in, or than long manure.

If buried too deeply, or put into a heavy soil, especially if in large quantity, it does not decay, but remains wet, and tends to make a bog of the field itself.

For soils that are rather heavy, it is therefore best to compost the peat with some rapidly fermenting manure. We thus get a compound which is quicker than muck, and slower than stable manure, etc., and is therefore better adapted to the wants of the soil than either of these would be alone.

Here it will be seen that much depends on the character of the peat itself. If light and spongy, and easily dried, it may be used alone with advantage on loamy soils, whereas if dense, and coherent, it would most likely be a poor amendment on a soil which has much tendency to become compact, and therefore does not readily free itself from excess of water.

But even a clay soil, if *thorough-drained and deeply plowed*, may be wonderfully improved by even a heavy dressing of muck, as then, the water being let off, the muck can exert no detrimental action; but operates as effectually to loosen a too heavy soil, as in case of sand, it makes an over-porous soil compact or retentive. A clay may be made friable, if well drained, by incorporating with it any substance as lime, sand, long manure or muck, which interposing between the clayey particles, prevents their adhering together.

II.—*Noxious ingredients.*

(*a*) *Vitriol peat.* Occasionally a peat is met with which is injurious if applied in the fresh state to crops, from its containing some substance which exerts a poisonous action on vegetation. The principal detrimental ingredients that occur in peat, appear to be sulphate of protoxide of iron,—the same body that is popularly known under the names copperas and green-vitriol,—and sulphate of alumina, the astringent component of alum.

I have found these substances ready formed in large quantity in but one of the peats that I have examined, viz.: that sent me by Mr. Perrin Scarborough, of Brooklyn, Conn. This peat dissolved in water to the extent of 15 *per cent.*, and the soluble portion, although containing some organic matter and sulphate of lime, consisted in great part of green-vitriol.

Portions of this muck, when thrown up to the air, become covered with "a white crust, having the taste of alum or saltpeter."

The bed containing this peat, though drained, yields but a little poor bog hay, and the peat itself, even after weathering for a year, when applied, mixed with one-fifth of stable manure to corn in the hill, gave no encouraging results, though a fair crop was obtained. It is probable that the sample analyzed was much richer in salts of iron and alumina, than the average of the muck.

Green-vitriol in minute doses is not hurtful, but rather beneficial to vegetation; but in larger quantity it is fatally destructive.

In a salt-marsh mud sent me by the Rev. Wm. Clift, of Stonington, Conn., there was found sulphate of iron in considerable quantity.

This noxious substance likewise occurred in small amount in swamp muck from E. Hoyt, Esq., New Canaan, Conn., and in hardly appreciable quantity in several others that I have examined. Besides green-vitriol, it is possible that certain organic salts of iron, may be deleterious.

The poisonous properties of vitriol-peats may be effectually corrected by composting with lime, or wood-ashes. By the action of these substances, sulphate of lime, (plaster of Paris) is formed, while the iron separates as peroxide, which, being insoluble, is without deleterious effect on vegetation. Where only soluble organic salts of iron (crenate of iron) are present, simple exposure to the air suffices to render them innocuous.

(*b*) *The acidity of Peats.* — Many writers have asserted that peat and muck possess a hurtful "acidity" which must be corrected before they can be usefully employed. It is indeed a fact, that peat consists largely of acids, but, except perhaps in the vitriol-peats, (those containing copperas,) they are so insoluble, or if soluble, are so quickly modified by the absorption of oxygen, that they do not exhibit any "acidity" that can be deleterious to vegetation. It is advised to neutralize this supposed acidity by lime or an alkali before using peat as a fertilizer or amendment, and there is great use in such mixtures of peat with alkaline matters, as we shall presently notice under the head of composts.

By the word acidity is conveyed the idea of something hurtful to plants. This something is, doubtless, in many cases, the salts of iron we have just noticed. In others, it is simply the inertness, "coldness" of the peat, which is not positively injurious, but is, for a time at least, of no benefit to the soil.

(*c*) ***Resinous matters*** are mentioned by various writers as injurious ingredients of peat, but I find no evidence that this notion is well-founded. The peat or muck formed from the decay of resinous wood and leaves does not appear to be injurious, and the amount of resin in peat is exceedingly small.

3.—*The Preparation of Peat for Agricultural use.*

(*a*) *Excavation.*—As to the time and manner of getting out peat, the circumstances of each case must determine. I only venture here to offer a few hints on this subject, which belongs so exclusively to the farm. The month of August is generally the appropriate time for throwing up peat, as then the swamps are usually most free from water, and most accessible to men and teams; but peat is often dug to best advantage in the winter, not only on account of the cheapness of labor, and from there being less hurry with other matters on the farm at that season, but also, because the freezing and thawing of the peat that is thrown out, greatly aid to disintegrate it and prepare it for use.

A correspondent of The *Homestead*, signing himself "Commentator," has given directions for getting out peat that are well worth the attention of farmers. He says: —

"The composting of muck and peat, with our stable and barn-yard manures, is surely destined to become one of the most important items in farm management throughout all the older States at least. One of the difficulties which lie in the way, is the first removal of the muck from its low and generally watery bed; to facilitate this, in many locations, it is less expensive to dry it before carting, by beginning an excavation at the border of the marsh in autumn, sufficiently wide for a cart path, throwing the muck out upon the surface on each side, and on a floor of boards or planks, to prevent it from absorbing moisture from the wet ground beneath; this broad ditch to be carried a sufficient length and depth to obtain the requisite quantity of muck. Thus thrown out, the two piles are now in a convenient form to be covered with boards, and, if properly done, the muck kept covered till the succeeding autumn, will be found to be dry and light, and in some cases may be carted away on the surface, or it may be best to let it remain a few months longer until the bottom of the ditch has become sufficiently frozen to bear a team; it can then be more easily loaded upon a sled or sleigh, and drawn to the yards and barn. In other localities, and where large quantities are wanted, and it lies deep, a sort of wooden railroad and inclined plane can be constructed by means of a plank track for the wheels of the cart to run upon, the team walking between these planks, and if the vehicle is inclined to 'run off the track,' it may usually be prevented by scantlings, say four inches thick, nailed upon one of the tracks on each side of the place where the wheel should run. Two or more teams and carts may now be employed, returning into the excavation outside of this track. As the work progresses, the track can be extended at both ends, and by continuing or increasing the inclination at the upper end, a large and high pile may be made, and if kept dry, will answer for years for composting, and can be easily drawn to the barn at any time."

(*b*) *Exposure, weathering, or seasoning of peat.*—In some cases, the chief or only use of exposing the thrown-up peat to the action of the air and weather during several months or a whole year, is to rid it of the great amount of water which adheres to it, and thus reduce its bulk and weight previous to cartage.

The general effect of exposure as indicated by my analyses, is to reduce the amount of matter soluble in water, and cause peats to approach in this respect a fertile soil, so that instead of containing 2, 4, or 6 *per cent.* of substances soluble in water, as at first, they are brought to contain but one-half these amounts, or even less. This change, however, goes on so rapidly after peat is mingled with the soil, that previous exposure on this account is rarely necessary, and most peats might be used perfectly fresh but for the difficulty often experienced, of reducing them to such a state of division as to admit of proper mixture with the soil.

The coherent peats which may be cut out in tough blocks, must be weathered, in order that the fibres of moss or grass-roots, which give them their consistency, may be decomposed or broken to an extent admitting of easy pulverization by the instruments of tillage.

The subjection of fresh and wet peat to frost, speedily destroys its coherence and reduces it to the proper state of pulverization. For this reason, fibrous peat should be exposed when wet to winter weather.

Another advantage of exposure is, to bring the peat into a state of more active chemical change. Peat, of the deeper denser sorts, is generally too inert ("sour," cold) to be directly useful to the plant. By exposure to the air it appears gradually to acquire the properties of the humus of the soil, or of stable manure, which are vegetable matters, altered by the same exposure. It appears to become more readily oxidable, more active, chemically, and thus more capable of exciting or rather aiding vegetable growth, which, so far as the soil is concerned, is the result of chemical activities.

Account has been already given of certain peats, which, used fresh, are accounted equal or nearly equal to stable manure. Others have come under the writer's notice, which have had little immediate effect when used before seasoning.

Mr. J. H. Stanwood says of a peat, from Colebrook, Conn., that it "has been used to some extent as a top-dressing for grass and other crops with satisfactory results, *although no particular benefit was noticeable during the first year*. After that, the effects might be seen for a number of years."

Rev. Wm. Clift observes, concerning a salt peat, from Stonington, Conn.: — "It has not been used fresh; is too acid; even potatoes do not yield well *in it the first season*, without manure."

The nature of the chemical changes induced by weathering, is to some extent understood so far as the nitrogen, the most important fertilizing element, is concerned. The nitrogen of peat, as we have seen, is mostly inert, a small portion of it only, existing in a soluble or available form. By weathering, portions of this nitrogen become converted into nitric acid. This action goes on at the surface of the heap, where it is most fully exposed to the air. Below, where the peat is more moist, ammonia is formed, perhaps simply by the reduction of nitric acid—not unlikely also, by the transformation of inert nitrogen. On referring to the analyses given on page 44, it is seen, that the first two samples contain but little ammonia and no nitric acid. Though it is not stated what was the condition of these peats, it is probable they had not been weathered. The other four samples were weathered, and the weathering had been the more effectual from the large admixture of sand with them. They yielded to the analyst very considerable quantities of ammonia and nitrates.

When a peat contains sulphate of protoxide of iron, or soluble organic salts of iron, to an injurious extent, these may be converted into other insoluble and innocuous bodies, by a sufficient exposure to the air. Sulphate of protoxide of iron is thus changed into sulphate of peroxide of iron, which is insoluble, and can therefore exert no hurtful effect on vegetation, while the soluble organic bodies of peat are oxydized and either converted into carbonic acid gas, carbonate of ammonia and water, or else made insoluble.

It is not probable, however, that merely throwing up a well characterized vitriol-peat into heaps, and exposing it thus imperfectly to the atmosphere, is sufficient to correct its bad qualities. Such peats need the addition of some alkaline body, as ammonia, lime, or potash, to render them salutary fertilizers.

(*c*) *This brings us to the subject of composting*, which appears to be the best means of taking full advantage of all the good qualities of peat, and of obviating or neutralizing the ill results that might follow the use of some raw peats, either from a peculiarity in their composition, (soluble organic compounds of iron, sulphate of protoxide of iron,) or from too great indestructibility. The chemical changes (oxidation of *iron* and *organic acids*), which prepare the inert or even hurtful ingredients of peat to minister to the support of vegetation, take place most rapidly in presence of certain other substances.

The subtances which rapidly induce chemical change in peats, are of two kinds, viz.: 1.—animal or vegetable matters that are highly susceptible to alteration and decay, and 2.—alkalies, either *ammonia* coming from the decomposition of animal matters, or *lime*, *potash* and *soda*.

A great variety of matters may of course be employed for making or mixing with peat composts; but there are comparatively few which allow of extensive and economical use, and our notice will be confined to these.

First of all, the composting of peat with *animal manures* deserves attention. Its advantages may be summed up in two statements.

1.—It is an easy and perfect method of economizing all such manures, even those kinds most liable to loss by fermentation, as night soil and horse-dung; and,

2.—It develops most fully and speedily the inert fertilizing qualities of the peat itself.

Without attempting any explanation of the changes undergone by a peat and manure compost, further than to say that the fermentation which begins in the manure extends to and involves the peat, reducing the whole nearly, if not exactly, to the condition of well-rotted dung, and that in this process the peat effectually prevents the loss of nitrogen as ammonia,—I may appropriately give the practical experience of farmers who have proved in the most conclusive manner how profitable it is to devote a share of time and labor to the manufacture of this kind of compost.

Preparation of Composts with Stable Manure.—The best plan of composting is to have a water tight trench, four inches deep and twenty inches wide, constructed in the stable floor, immediately behind the cattle, and every morning put a bushel-basketful of muck behind each animal. In this way the urine is perfectly absorbed by the muck, while the warmth of the freshly voided excrements so facilitates the fermentative process, that, according to Mr. F. Holbrook, Brattleboro, Vt., who has described this method, *much more muck can thus be well prepared for use* in the spring, than by any of the ordinary modes of composting. When the dung and muck are removed from the stable, they should be well intermixed, and as fast as the compost is prepared, it should be put into a compact heap, and covered with a layer of muck several inches thick. It will then hardly require any shelter if used in the spring.

If the peat be sufficiently dry and powdery, or free from tough lumps, it may usefully serve as bedding, or litter for horses and cattle, as it absorbs the urine, and is sufficiently mixed with the dung in the operation of cleaning the stable. It is especially good in the pig-pen, where the animals themselves work over the compost in the most thorough manner, especially if a few kernels of corn be occasionally scattered upon it.

Mr. Edwin Hoyt, of New Canaan, Conn., writes:—"Our horse stables are constructed with a movable floor and pit beneath, which holds 20 loads of muck of 25 bushels per load. Spring and fall, this pit is filled with fresh muck, which receives all the urine of the horses, and being occasionally worked over and mixed, furnishes us annually with 40 loads of the most valuable manure."

"Our stables are sprinkled with muck every morning, at the rate of one bushel per stall, and the smell of ammonia, etc., so offensive in most stables, is never perceived in ours. Not only are the stables kept sweet, but the ammonia is saved by this procedure."

When it is preferred to make the compost out of doors, the plan generally followed is to lay down a bed of weathered peat, say eight to twelve inches thick; cover this with a layer of stable dung, of four to eight inches; put on another stratum of peat, and so, until a heap of three to four feet is built up. The heap may be six to eight feet wide, and indefinitely long. It should be finished with a thick coating of peat, and the manure should be covered as fast as brought out.

The proportions of manure and peat should vary somewhat according to their quality and characters. Strawy manure, or that from milch-cows, will "ferment" less peat than clear dung, especially when the latter is made by horses or highly fed animals. Some kinds of peat heat much easier than others. There are peats which will ferment of themselves in warm moist weather—even in the bog, giving off ammonia in perceptible though small amount. Experience is the only certain guide as to the relative quantities to be employed, various proportions from one to five of peat for one of manure, by bulk, being used.

When the land is light and needs amending, as regards its retentive power, it is best to make the quantity of peat as large as can be thoroughly fermented by the manure.

The making of a high heap, and the keeping it trim and in shape, is a matter requiring more labor than is generally necessary. Mr. J. H. Stanwood, of Colebrook, Conn., writes me: —

"My method of composting is as follows: I draw my muck to the barn-yard, placing the loads as near together as I can tip them from the cart. Upon this I spread whatever manure I have at hand, and mix with the feet of the cattle, and heap up with a scraper."

Peat may be advantageously used to save from waste the droppings of the yard.

Mr. Edwin Hoyt, of New Canaan, Conn., says:— "We use muck largely in our barn-yards, and after it becomes thoroughly saturated and intermixed with the droppings of the stock, it is piled up to ferment, and the yard is covered again with fresh muck."

Mr. N. Hart, Jr., of West Cornwall, Conn., writes: — "In the use of muck we proceed as follows: Soon after haying we throw up enough for a year's use, or several hundred loads. In the fall, the summer's accumulation in hog-pens and barn cellars is spread upon the mowing grounds, and a liberal supply of muck carted in and spread in the bottoms of the cellars, ready for the season for stabling cattle. When this is well saturated with the drippings of the stables, a new supply is added. The accumulation of the winter is usually applied to the land for the corn crop, except the finer portion, which is used to top-dress meadow land. A new supply is then drawn in for the swine to work up. This is added to from time to time, and as the swine are fed on whey, they will convert a large quantity into valuable manure for top-dressing mowing land."

A difference of opinion exists as to the treatment of the compost. Some hold it indifferent whether the peat and manure are mixed, or put in layers, when the composting begins. Others assert, that the fermentation proceeds better when the ingredients are stratified. Some direct, that the compost should not be stirred. The general testimony is, that mixture, at the outset, is as effectual as putting up in layers; but, if the manure be strawy, it is, of course, difficult or impracticable to mix at first. Opinion also preponderates in favor of stirring, during or after the fermentation.

Mr. Hoyt remarks:—"We are convinced, that the oftener a compost pile of yard manure and muck is worked over after fermenting, the better. We work it over and add to it a little more muck and other material, and the air being thus allowed to penetrate it, a new fermentation or heating takes place, rendering it more decomposable and valuable."

Rev. Wm. Clift, writes:—"Three or four loads of muck to one of stable manure, put together in the fall or winter in alternate layers, forked over twice before spreading and plowing in, may represent the method of composting."

Mr. Adams White, of Brooklyn, Conn., proceeds in a different manner. He says:—"In composting, 20 loads are drawn on to upland in September, and thrown up in a long pile. Early in the spring 20 loads of stable manure are laid along side, and covered with the muck. As soon as it has heated moderately, the whole is forked over and well mixed."

Those who have practiced making peat composts with their yard, stable, and pen manure, almost invariably find them highly satisfactory in use, especially upon light soils.

A number of years ago, I saw a large pile of compost in the farm-yard of Mr. Pond, of Milford, Conn., and witnessed its effect as applied by that gentleman to a field of sixteen acres of fine gravelly or coarse sandy soil. The soil, from having a light color and excessive porosity, had become dark, unctuous, and retentive of moisture, so that during the drouth of 1856, the crops on this field were good and continued to flourish, while on the contiguous land they were dried up and nearly ruined. This compost was made from a light muck, that contained but three *per cent.* of ash (more than half of which was sand), and but 1.2 *per cent.* of nitrogen, in the air-dry state — (twenty *per cent.* of water). Three loads of this muck were used to one of stable manure.

Here follow some est'mates of the value of this compost by practical men. They are given to show that older statements, to the same effect, cannot be regarded as exaggerated.

Mr. J. H. Stanwood, of Colebrook, Conn., says: — "Experiments made by myself, have confirmed me in the opinion that a compost of equal parts of muck and stable manure is equal to the same quantity of stable manure."

Mr. Daniel Buck, Jr., of Poquonnock, Conn., remarks: — "8 loads of muck and 4 of manure in compost, when properly forked over, are equal to 12 loads of barn-yard manure on sandy soil."

Rev. Wm. Clift, of Stonington, Conn., writes: — "I consider a compost made of one load of stable manure and three of muck, equal in value to four loads of yard manure."

Mr. N. Hart, Jr., of West Cornwall, Conn., observes of a peat sent by him for analysis: — "We formerly composted it in the yard with stable manure, but have remodeled our stables, and now use it as an absorbent and to increase the bulk of manure to double its original quantity. We consider the mixture more valuable than the same quantity of stable manure." Again, "so successful has been the use of it, that we could hardly carry on our farming operations without it."

Mr. Adams White, of Brooklyn, Conn., states:— "The compost of equal bulks of muck and stable manure, has been used for corn (with plaster in the hill,) on dry sandy soil to great advantage. I consider the compost worth more per cord than the barn-yard manure."

Night Soil is a substance which possesses, when fresh, the most valuable fertilizing qualities, in a very concentrated form. It is also one which is liable to rapid and almost complete deterioration, as I have demonstrated by analyses. The only methods of getting the full effect of this material are, either to use it fresh, as is done by the Chinese and Japanese on a most extensive and offensive scale; or to compost it before it can decompose. The former method, will, it is to be hoped, never find acceptance among us. The latter plan has nearly all the advantages of the former, without its unpleasant features.

When the night soil falls into a vault, it may be composted, by simply sprinkling fine peat over its surface, once or twice weekly, as the case may require, *i. e.* as often as a bad odor prevails. The quantity thus added, may be from twice to ten times the bulk of the night soil, —the more within these limits, the better. When the vault is full, the mass should be removed, worked well over and after a few days standing, will be ready to use to manure corn, tobacco, etc., in the hill, or for any purpose to which guano or poudrette is applied. If it cannot be shortly used, it should be made into a compact heap, and covered with a thick stratum of peat. When signs of heating appear, it should be watched closely; and if the process attains too much violence, additional peat should be worked into it. Drenching with water is one of the readiest means of checking too much heating, but acts only temporarily. Dilution with peat to a proper point, which experience alone can teach, is the surest way of preventing loss. It should not be forgotten to put a thick layer of peat at the bottom of the vault to begin with.

Another excellent plan, when circumstances admit, is, to have the earth-floor where the night soil drops, level with the surface of the ground, or but slightly excavated, and a shed attached to the rear of the privy to shelter a good supply of peat as well as the compost itself. Operations are begun by putting down a layer of peat to receive the droppings; enough should be used to absorb all the urine. When this is nearly saturated, more should be sprinkled on, and the process is repeated until the accumulations must be removed to make room for more. Then, once a week or so, the whole is hauled out into the shed, well mixed, and formed into a compact heap, or placed as a layer upon a stratum of peat, some inches thick, and covered with the same. The quantity of first-class compost that may be made yearly upon any farm, if due care be taken, would astonish those who have not tried it. James Smith, of Deanston, Scotland, who originated our present system of Thorough Drainage, asserted, that the excrements of one man for a year, are sufficient to manure half an acre of land. In Belgium the manure from such a source has a commercial value of $9.00 gold.

It is certain, that the skillful farmer may make considerably more than that sum from it in New England, *per annum*. Mr. Hoyt, of New Canaan, Conn., says: —

"Our privies are deodorized by the use of muck, which is sprinkled over the surface of the pit once a week, and from them alone we thus prepare annually, enough "poudrette" to manure our corn in the hill."

Peruvian Guano, so serviceable in its first applications to light soils, may be composted with muck to the greatest advantage. Guano is an excellent material for bringing muck into good condition, and on the other hand the muck most effectually prevents any waste of the costly guano, and at the same time, by furnishing the soil with its own ingredients, to a greater or less degree prevents the exhaustion that often follows the use of guano alone. The quantity of muck should be pretty large compared to that of the guano,—a bushel of guano will compost six, eight, or ten of muck. Both should be quite fine, and should be well mixed, the mixture should be moist and kept covered with a layer of muck of several inches of thickness. This sort of compost would probably be sufficiently fermented in a week or two of warm weather, and should be made and kept under cover.

If no more than five or six parts of muck to one of guano are employed, the compost, according to the experience of Simon Brown, Esq., of the Boston *Cultivator*, (Patent Office Report for 1856), will prove injurious, if placed in the hill in contact with seed, but may be applied broadcast without danger.

The ***Menhaden***, or "***White fish***", so abundantly caught along our Sound coast during the summer months, or any variety of fish may be composted with muck, so as to make a powerful manure, with avoidance of the excessively disagreeable stench which is produced when these fish are put directly on the land. Messrs. Stephen Hoyt & Sons, of New Canaan, Conn., make this compost on a large scale. I cannot do better than to give entire Mr. Edwin Hoyt's account of their operations, communicated to me several years ago.

"During the present season, (1858,) we have composted about 200,000 white fish with about 700 loads (17,500 bushels) of muck. We vary the proportions somewhat according to the crop the compost is intended for. For rye we apply 20 to 25 loads per acre of a compost made with 4,500 fish, (one load) and with this manuring, no matter how poor the soil, the rye will be as large as a man can cradle. Much of ours we have to reap. For oats we use less fish, as this crop is apt to lodge. For corn, one part fish to ten or twelve muck is about right, while for grass or any top-dressing, the proportion of fish may be increased."

"We find it is best to mix the fish in the summer and not use the compost until the next spring and summer. Yet we are obliged to use in September for our winter rye a great deal of the compost made in July. We usually compost the first arrivals of fish in June for our winter grain; after this pile has stood three or four weeks, it is worked over thoroughly. In this space of time the fish become pretty well decomposed, though they still preserve their form and smell outrageously. As the pile is worked over, a sprinkling of muck or plaster is given to retain any escaping ammonia. At the time of use in September the fish have completely disappeared, bones and fins excepted."

"The effect on the muck is to blacken it and make it more loose and crumbly. As to the results of the use of this compost, we find them in the highest degree satisfactory. We have raised 30 to 35 bushels of rye per acre on land that without it could have yielded 6 or 8 bushels at the utmost. This year we have corn that will give 60 to 70 bushels per acre, that otherwise would yield but 20 to 25 bushels. It makes large potatoes, excellent turnips and carrots."

Fish compost thus prepared, is a uniform mass of fishy but not putrefactive odor, not disagreeable to handle. It retains perfectly all the fertilizing power of the fish. Lands, manured with this compost, will keep in heart and improve: while, as is well known to our coast farmers, the use of fish alone is ruinous in the end, on light soils.

It is obvious that *any other easily decomposing animal matters, as slaughter-house offal, soap boiler's scraps, glue waste, horn shavings, shoddy, castor pummace, cotton seed-meal, etc., etc.*, may be composted in a similar manner, and that several or all these substances may be made together into one compost.

In case of the composts with yard manure, guano and other animal matters, the alkali, *ammonia*, formed in the fermentation, greatly promotes chemical change, and it would appear that this substance, on some accounts, excels all others in its efficacy. The other alkaline bodies, *potash*, *soda* and *lime*, are however scarcely less active in this respect, and being at the same time, of themselves, useful fertilizers, they also may be employed in preparing muck composts.

Potash-lye and *soda-ash* have been recommended for composting with muck; but, although they are no doubt highly efficacious, they are too costly for extended use.

The other alkaline materials that may be cheaply employed, and are recommended, are *wood-ashes*, leached and unleached, *ashes of peat*, *shell marl*, (consisting of carbonate of lime,) *quick lime*, *gas lime*, and what is called "*salt and lime mixture*."

With regard to the proportions to be used, no very definite rules can be laid down; but we may safely follow those who have had experience in the matter. Thus, to a cord of muck, which is about 100 bushels, may be added, of unleached wood ashes twelve bushels, or of leached wood ashes twenty bushels, or of peat ashes twenty bushels, or of marl, or of gas lime twenty bushels. Ten bushels of quick lime, slaked with water or salt-brine previous to use, is enough for a cord of muck.

Instead of using the above mentioned substances singly, any or all of them may be employed together.

The muck should be as fine and free from lumps as possible, and must be intimately mixed with the other ingredients by shoveling over. The mass is then thrown up into a compact heap, which may be four feet high. When the heap is formed, it is well to pour on as much water as the mass will absorb, (this may be omitted if the muck is already quite moist,) and finally the whole is covered over with a few inches of pure muck, so as to retain moisture and heat. If the heap is put up in the Spring, it may stand undisturbed for one or two months, when it is well to shovel it over and mix it thoroughly. It should then be built up again, covered with fresh muck, and allowed to stand as before until thoroughly decomposed. The time required for this purpose varies with the kind of muck, and the quality of the other material used. The weather and thoroughness of intermixture of the ingredients also materially affect the rapidity of decomposition. In all cases five or six months of summer weather is a sufficient time to fit these composts for application to the soil.

Mr. Stanwood of Colebrook, Conn., says: "I have found a compost made of two bushels of unleached ashes to twenty-five of muck, superior to stable manure as a top-dressing for grass, on a warm, dry soil."

N. Hart, Jr., of West Cornwall, Conn., states: "I have mixed 25 bushels of ashes with the same number of loads of muck, and applied it to ¾ of an acre. The result was far beyond that obtained by applying 300 lbs. best guano to the same piece."

The use of "*salt and lime mixture*" is so strongly recommended, that a few words may be devoted to its consideration.

When quick-lime is slaked with a brine of common salt (chloride of sodium), there are formed by double decomposition, small portions of caustic soda and chloride of calcium, which dissolve in the liquid. If the solution stand awhile, carbonic acid is absorbed from the air, forming carbonate of soda: but carbonate of soda and chloride of calcium instantly exchange their ingredients, forming insoluble carbonate of lime and reproducing common salt.

When the fresh mixture of quick-lime and salt is incorporated with *any porous body*, as soil or peat, then, as Graham has shown, *unequal diffusion* of the caustic soda and chloride of calcium occurs from the point where they are formed, through the moist porous mass, and the result is, that the small portion of caustic soda which diffuses most rapidly, or the carbonate of soda formed by its speedy union with carbonic acid, is removed from contact with the chloride of calcium.

Soda and carbonate of soda are more soluble in water and more strongly alkaline than lime. They, therefore, act on peat more energetically than the latter. It is on account of the formation of soda and carbonate of soda from the lime and salt mixture, that this mixture exerts a more powerful decomposing action than lime alone. Where salt is cheap and wood ashes scarce, the mixture may be employed accordingly to advantage. Of its usefulness we have the testimony of practical men.

Says Mr. F. Holbrook of Vermont, (Patent Office Report for 1856, page 193.) "I had a heap of seventy-five half cords of muck mixed with lime in the proportion of a half cord of muck to a bushel of lime. The muck was drawn to the field when wanted in August. A bushel of salt to six bushels of lime was dissolved in water enough to slake the lime down to a fine dry powder, the lime being slaked no faster than wanted, and spread immediately while warm, over the layers of muck, which were about six inches thick; then a coating of lime and so on, until the heap reached the height of five feet, a convenient width, and length enough to embrace the whole quantity of the muck. In about three weeks a powerful decomposition was apparent, and the heap was nicely overhauled, nothing more being done to it till it was loaded the next Spring for spreading. The compost was spread on the plowed surface of a dry sandy loam at the rate of about fifteen cords to the acre, and harrowed in. The land was planted with corn and the crop was more than sixty bushels to the acre."

Other writers assert that they "have decomposed with this mixture, spent tan, saw dust, corn stalks, swamp muck, leaves from the woods, indeed every variety of inert substance, and in *much shorter time than it could be done by any other means.*" (Working Farmer, Vol. III. p. 280.)

Some experiments that have a bearing on the efficacy of this compost will be detailed presently.

There is no doubt that the soluble and more active (caustic) forms of alkaline bodies exert a powerful decomposing and solvent action on peat. It is asserted too that the *nearly insoluble and less active matters of this kind,* also have an effect, though a less complete and rapid one. Thus, *carbonate of lime* in the various forms of chalk, shell marl,* old mortar, leached ashes and peat ashes, (for in all these it is the chief and most "alkaline" in-

* *Shell marl,* consisting of fragments and powder of fresh-water shells, is frequently met with, underlying peat beds. Such a deposit occurs on the farm of Mr. John Adams, in Salisbury, Conn. It is eight to ten feet thick. An air-dry sample, analyzed under the writer's direction, gave results as follows:

Water			30.62
Organic matter	soluble in water	0.70	6.52
	insoluble in water	5.82	
Carbonate of lime			57.09
Sand			1.86
Oxide of iron and alumina, with traces of potash, magnesia, sulphuric and phosphoric acid			3.91
			100.00

Another specimen from near Milwaukee, Wis., said to occur there in immense quantities underlying peat, contained, by the author's analysis —

Water	1.14
Carbonate of lime	92.41
Carbonate of magnesia	3.43
Peroxide of iron with a trace of phosphoric acid	0.92
Sand	1.60
	99.50

gredient,) is recommended to compost with peat. Let us inquire whether carbonate of lime can really exert any noticeable influence in improving the fertilizing quality of peat.

In the case of vitriol peats, carbonate of lime is the cheapest and most appropriate means of destroying the noxious sulphate of protoxide of iron, and correcting their deleterious quality. When carbonate of lime is brought in contact with sulphate of protoxide of iron, the two bodies mutually decompose, with formation of sulphate of lime (gypsum) and carbonate of protoxide of iron. The latter substance absorbs oxygen from the air with the utmost avidity, and passes into the peroxide of iron, which is entirely inert.

The admixture of any earthy matter with peat, will facilitate its decomposition, and make it more active chemically, in so far as it promotes the separation of the particles of the peat from each other, and the consequent access of air. This benefit may well amount to something when we add to peat one-fifth of its bulk of marl or leached ashes, but the question comes up: Do these insoluble mild alkalies exert any direct action? Would not as much soil of any kind be equally efficacious, by promoting to an equal degree the contact of oxygen from the atmosphere?

There are two ways in which carbonate of lime may exert a chemical action on the organic matters of peat. Carbonate of lime, itself, in the forms we have mentioned, is commonly called insoluble in water. It is, however, soluble to a very slight extent; it dissolves, namely, in about 30,000 times its weight of pure water. It is nearly thirty times more soluble in water saturated with carbonic acid; and this solution has distinct alkaline characters. Since the water contained in a heap of peat must be considerably impregnated with carbonic acid, it follows that when carbonate of lime is present, the latter must form a solution, very dilute indeed, but still capable of some direct effect on the organic matters of the peat, when it acts through a long space of time. Again, it is possible that the solution of carbonate of lime in carbonic acid, may act to liberate some ammonia from the soluble portions of the peat, and this ammonia may react on the remainder of the peat to produce the same effects as it does in the case of a compost made with animal matters.

Whether the effects thus theoretically possible, amount to anything practically important, is a question of great interest. It often happens that opinions entertained by practical men, not only by farmers, but by mechanics and artisans as well, are founded on so untrustworthy a basis, are supported by trials so destitute of precision, that their accuracy may well be doubted, and from all the accounts I have met with, it does not seem to have been well established, practically, that composts made with carbonate of lime, are better than the peat and carbonate used separately.

Carbonate of lime (leached ashes, shell marl, etc.), is very well to use *in conjunction with* peat, to furnish a substance or substances needful to the growth of plants, and supply the deficiencies of peat as regards composition. Although in the agricultural papers, numerous accounts of the efficacy of such mixtures are given, we do not learn from them whether these bodies exert any such good effect upon the peat itself, as to warrant the trouble of making a *compost.*

4.—*Experiments by the author on the effect of alkaline bodies in developing the fertilizing power of Peat.*

During the summer of 1862, the author undertook a series of experiments with a view of ascertaining the effect of various composting materials upon peat.

Two bushels of peat were obtained from a heap that had been weathering for some time on the "Beaver Meadow," near New Haven. This was thoroughly air-dried, then crushed by the hand, and finally rubbed through a moderately fine sieve. In this way, the peat was brought to a perfectly homogeneous condition.

Twelve-quart flower-pots, new from the warehouse, were filled as described below; the trials being made in duplicate: —

Pots 1 and 2 contained each 270 grammes of peat.

Pots 3 and 4 contained each 270 grammes of peat, mixed with 10 grammes of ashes of young grass.

Pots 5 and 6 contained each 270 grammes of peat, 10 grammes of ashes, and 10 grammes of carbonate of lime.

Pots 7 and 8 contained each 270 grammes of peat, 10 grammes of ashes, and 10 grammes of slaked (hydrate of) lime.

Pots 9 and 10 contained each 270 grammes of peat, 10 grammes of ashes, and 5 grammes of lime, slaked with strong solution of common salt.

Pots 11 and 12 contained each 270 grammes of peat, 10 grammes of ashes, and 3 grammes of Peruvian guano.

In each case the materials were thoroughly mixed together, and so much water was cautiously added as served to wet them thoroughly. Five kernels of dwarf (pop) corn were planted in each pot, the weight of each planting being carefully ascertained.

The pots were disposed in a glazed case within a cold grapery,* and were watered when needful with pure water. The seeds sprouted duly, and developed into healthy plants. The plants served thus as tests of the

* To the kindness of Joseph Sheffield, Esq., of New Haven, the author is indebted for facilities in carrying on these experiments.

chemical effect of carbonate of lime, of slaked lime, and of salt and lime mixture, on the peat. The guano pots enabled making a comparison with a well-known fertilizer. The plants were allowed to grow until those best developed, enlarged above, not at the expense of the peat, etc., but of their own lower leaves, as shown by the withering of the latter. They were then cut, and, after drying in the air, were weighed with the subjoined results.

VEGETATION EXPERIMENTS IN PEAT COMPOSTS.

Nos.	*Medium of Growth.*	*Weight of crops in grammes.*		*Comparative weight of crops, the sum of 1. and 2. taken as unity.*	*Ratio of weight of crops to weight of seeds, the latter assumed as unity.*
1 2	Peat alone.	1.61 2.59	4.20	1	2½
3 4	Peat, and ashes of grass,	14.19 18.25	32.44	8	20½
5 6	Peat, ashes, and carbonate of lime,	18.19 20.25	38.44	9	25½
7 8	Peat, ashes, and slaked lime,	21.49 20.73	42.22	10	28½
9 10	Peat, ashes, slaked lime, and salt,	23.08 23.34	46.42	11	30½
11 12	Peat, ashes, and Peruvian Guano,	26.79 26.99	53.78	13	35½

Let us now examine the above results. The experiments 1 and 2, demonstrate that the peat itself is deficient in something needful to the plant. In both pots, but 4.2 grammes of crop were produced, a quantity two and a half times greater than that of the seeds, which weighed 1.59 grammes. The plants were pale in color, slender, and reached a height of but about six inches.

Nos. 3 and 4 make evident what are some of the deficiencies of the peat. A supply of mineral matters, such as are contained in all plants, being made by the addition of *ashes*, consisting chiefly of phosphates, carbonates and sulphates of lime, magnesia and potash, a crop is realized nearly eight times greater than in the previous cases; the yield being 32.44 grammes, or 20½ times the weight of the seed. The quantity of ashes added, viz.: — 10 grammes, was capable of supplying every mineral element, greatly in excess of the wants of any crop that could be grown in a quart of soil. The plants in pots 3 and 4 were much stouter than those in 1 and 2, and had a healthy color.

The experiments 5 and 6 appear to demonstrate that *carbonate of lime* considerably aided in converting the peat itself into plant-food. The ashes alone contained enough carbonate of lime to supply the wants of the plant in respect to that substance. More carbonate of lime could only operate by acting on the organic matters of the peat. The amount of the crop is raised by the effect of carbonate of lime from 32.44 to 38.44 grammes, or from 20½ to 25½ times that of the seed.

Experiments 7 and 8 show, that *slaked lime* has more effect than the carbonate, as we should anticipate. Its influence does not, however, exceed that of the carbonate very greatly, the yield rising from 38.44 to 42.22 grammes, or from 25½ to 28½ times the weight of the seed. In fact, quicklime can only act as such for a very short space of time, since it rapidly combines with the carbonic acid, which is supplied abundantly by the peat. In experiments 7 and 8, a good share of the influence exerted must therefore be actually ascribed to the carbonate, rather than to the quicklime itself.

In experiments 9 and 10, we have proof that the "*lime and salt mixture*" has a greater efficacy than lime alone, the crop being increased thereby from 42.22 to 46.42 grammes, or from 28½ to 30½ times that of the seed.

Finally, we see from experiments 11 and 12 that in all the foregoing cases it was a limited supply of *nitrogen* that limited the crop; for, on adding Peruvian guano, which could only act by this element (its other ingredients, phosphates of lime and potash, being abundantly supplied in the ashes), the yield was carried up to 53.78 grammes, or $35\frac{1}{2}$ times the weight of the seed, and 13 times the weight of the crop obtained from the unmixed peat.

5.—*The Examination of Peat* (*muck and marsh-mud*) *with reference to its Agricultural Value.*

Since, as we are forced to conclude, the variations in the composition of peat stand in no recognizable relations to differences of appearance, it is only possible to ascertain the value of any given specimen by actual trial or by chemical investigation.

The method *by practical trial* is usually the cheaper and more satisfactory of the two, though a half year or more is needful to gain the desired information.

It is sufficient to apply to small measured plots of ground, each say two rods square, known quantities of the fresh, the weathered, and the composted peat in order, by comparison of the growth and *weight* of the crop, to decide the question of their value.

Peat and its composts are usually applied at rates ranging from 20 to 40 wagon or cart loads per acre. There being 160 square rods in the acre, the quantity proper to a plot of two rods square (= four square rods,) would be one half to one load.

The composts with stable manure and lime, or salt and lime mixture, are those which, in general, it would be best to experiment with. From the effects of the stable manure compost, could be inferred with safety the value of any compost, of which animal manure is an essential ingredient.

One great advantage of the practical trial on the small scale is, that the adaptation of the peat or of the compost to the *peculiarities of the soil*, is decided beyond a question.

It must be borne in mind, however, that the results of experiments can only be relied upon, when the plots are accurately measured, when the peat, etc., are applied in known quantities, and when the crops are separately harvested and carefully weighed.

If experiments are made upon grass or clover, the gravest errors may arise by drawing conclusions from the appearance of the standing crop. Experience has shown that two clover crops, gathered from contiguous plots differently manured, may strikingly differ in appearance, but yield the same amounts of hay.

The *chemical examination* of a peat may serve to inform us, without loss of time, upon a number of important points.

To test a peat for *soluble iron salts* which might render it deleterious, we soak and agitate a handful for some hours, with four or five times its bulk of warm soft water. From a *good fresh-water peat* we obtain, by this treatment, a yellow liquid, more or less deep in tint, the taste of which is very slight and scarcely definable.

From a *vitriol peat* we get a dark-brown or black solution, which has a bitter, astringent, metallic or inky taste, like that of copperas.

Salt peat will yield a solution having the taste of salt-brine, unless it contains iron, when the taste of the latter will prevail.

On evaporating the water-solution to dryness and heating strongly in a China cup, a *vitriol peat* gives off white choking fumes of sulphuric acid, and there remains, after burning, brown-red oxide of iron in the dish.

The above testings are easily conducted by any one, with the ordinary conveniences of the kitchen.

Those that follow, require, for the most part, the chemical laboratory, and the skill of the practised chemist, for satisfactory execution.

Besides testing for soluble iron compounds, as already indicated, the points to be regarded in the chemical examination, are: —

1st. *Water or moisture.*—This must be estimated, because it is so variable, and a knowledge of its quantity is needful, if we will compare together different samples. A weighed amount of the peat is dried for this purpose at 212° F., as long as it suffers loss.

2d. The *proportions of organic matter and ash* are ascertained by carefully burning a weighed sample of the peat. By this trial we distinguish between peat with 2 to 10 *per cent.* of ash and peaty soil, or mud, containing but a few *per cent.* of organic matter.

This experiment may be made in a rough way, but with sufficient accuracy for common purposes, by burning a few lbs. or ozs. of peat upon a piece of sheet iron, or in a sauce pan, and noting the loss, which includes both *water* and *organic matter.*

3d. As further regards the organic matters, we ascertain *the extent to which the peaty decomposition has taken place* by boiling with dilute solution of carbonate of soda. This solvent separates the humic and ulmic acids from the undecomposed vegetable fibers.

For practical purposes this treatment with carbonate of soda may be dispensed with, since the amount of undecomposed fiber is gathered with sufficient accuracy from careful inspection of the peat.

Special examination of the organic acids is of no consequence in the present state of our knowledge.

4th. The *proportion of nitrogen* is of the first importance to be ascertained. In examinations of 30 samples of peat, I have found the content of nitrogen to range from 0.4 to 2.9 *per cent.*, the richest containing seven times as much as the poorest. It is practically a matter of great moment whether, for example, a Peruvian guano contains 16 *per cent.* of nitrogen as it should, or but one-seventh that amount, as it may when grossly adulterated. In the same sense, it is important before making a heavy outlay in excavating and composting peat, to know whether (as regards nitrogen) it belongs to the poorer or richer sorts. This can only be done by the complicated methods known to the chemist.

5th. The estimation of *ammonia* (actual or ready-formed,) is a matter of scientific interest, but subordinate in a practical point of view.

6th. *Nitric acid* and *nitrates* can scarcely exist in peat except where it is well exposed to the air, in a merely moist but not wet state. Their estimation in composts is of great interest, though troublesome to execute.

7th. As regards the ash, its red color indicates *iron.* Pouring hydrochloric acid upon it, causes effervescence in the presence of *carbonate of lime.* This compound, in most cases, has been formed in the burning, from humate and other organic salts of lime. *Sand,* or *clay,* being insoluble in the acid, remains, and may be readily estimated.

Phosphoric acid and alkalies, especially *potash,* are, next to lime, the important ingredients of the ash. *Magnesia* and *sulphuric acid,* rank next in value. Their estimation requires a number of tedious operations, and can scarcely be required for practical purposes, until more ready methods of analyses shall have been discovered.

8th. The quantity of *matters soluble in water* has considerable interest, but is not ordinarily requisite to be ascertained.

6.—*Composition of Connecticut Peats.*

In the years 1857 and 1858, the author was charged by the Connecticut State Agricultural Society* with the

* At the instigation of Henry A Dyer, Esq., at that time the Society's Corresponding Secretary.

chemical investigation of 33 samples of peat and swamp muck, sent to him in compliance with official request.

In the foregoing pages, the facts revealed by the laborious analyses executed on these samples, have been for the most part communicated, together with many valuable practical results derived from the experience of the gentlemen who sent in the specimens. The analytical data themselves appear to me to be worthy of printing again, for the information of those who may hereafter make investigations in the same direction.—See Tables I, II, and III, p.p. 89, 90, and 91.

The specimens came in all stages of dryness. Some were freshly dug and wet, others had suffered long exposure, so that they were air-dry; some that were sent in the moist state, became dry before being subjected to examination; others were prepared for analysis while still moist.

A sufficient quantity of each specimen was carefully pulverized, intermixed, and put into a stoppered bottle and thus preserved for experiment.

The analyses were begun in the winter of 1857 by my assistant, Edward H. Twining, Esq. The samples 1 to 17 of the subjoined tables were then analyzed. In the following year the work was continued on the remaining specimens 18—33 by Dr. Robert A. Fisher. The method of analysis was the same in both cases, except in two particulars.

In the earlier analyses, 1 to 17 inclusive, the treatment with carbonate of soda was not carried far enough to dissolve the whole of the soluble organic acids. It was merely attempted to make *comparative* determinations by treating all alike for the same time, and with the same quantity of alkali. I have little doubt that in some cases not more than one-half of the portion really soluble in carbonate of soda is given as such. In the later analyses,

18 to 33, however, the treatment was continued until complete separation of the soluble organic acids was effected.

By acting on a peat for a long time with a hot solution of carbonate of soda, there is taken up not merely a quantity of organic matter, but inorganic matters likewise enter solution. Silica, oxyd of iron and alumina are thus dissolved. In this process too, sulphate of lime is converted into carbonate of lime.

The total amount of these soluble inorganic matters has been determined with approximate accuracy in analyses 18 to 33.

In the analyses 1 to 17 the collective amount of matters soluble in water was determined. In the later analyses the proportions of organic and inorganic matters in the water-solution were separately estimated.

The process of analysis as elaborated and employed by Dr. Fisher and the author, is as follows:

I. To prepare a sample for analysis, half a pound, more or less, of the substance is pulverized and passed through a wire sieve of 24 meshes to the inch. It is then thoroughly mixed and bottled.

II. 2 grammes of the above are dried (in tared watch-glasses) at the temperature of 212 degrees, until they no longer decrease in weight. The loss sustained represents the *amount of water*, (according to MARSILLY, Annales des Mines, 1857, XII., 404, peat loses carbon if dried at a temperature higher than 212 degrees.)

III. The capsule containing the residue from I. is slowly heated to incipient redness, and maintained at that temperature until the organic matter is entirely consumed. The loss gives the total amount of *organic*, the residue the total amount of *inorganic* matter.

NOTE.—In peats containing sulphate of the protoxide of iron, the loss that occurs during ignition is partly due to the escape of sulphuric acid, which is set free by the decomposition of the above mentioned salt of iron. But the quantity is usually so small in comparison with the organic matter, that it may be disregarded. The same may be said of the combined water in the clay that is mixed with some mucks, which is only expelled at a high temperature.

IV. 3 grammes of the sample are digested for half an hour, with 200 cubic centimeters (66.6 times their weight,) of boiling water, then removed from the sand bath, and at the end of twenty-four hours, the clear liquid is decanted. This operation is twice repeated upon the residue; the three solutions are mixed, filtered, concentrated, and finally evaporated to dryness (in a tared platinum capsule,) over a water bath. The residue, which must be dried at 212 degrees, until it ceases to lose weight, gives the *total amount soluble in water*. The dried residue is then heated to low redness, and maintained at that temperature until the organic matter is burned off. The loss represents the amount of *organic matter soluble in water*, the ash gives the quantity of *soluble inorganic matter*.

V. 1 gramme is digeste l for two hours, at a temperature just below the boiling point, with 100 cubic centimeters of a solution containing 5 *per cent.* of crystallized carbonate of soda. It is then removed from the sand bath and allowed to settle. When the supernatant liquid has become perfectly transparent, it is carefully decanted. This operation is repeated until all the organic matter soluble in this menstruum is removed; which is accomplished as soon as the carbonate of soda solution comes off colorless. The residue, which is to be washed with boiling water until the washings no longer affect test papers, is thrown upon a tared filter, and dried at 212 degrees. It is the *total amount of organic and inorganic matter insoluble in carbonate of soda.* The loss that it suffers upon ignition, indicates the amount of *organic matter*, the ash gives the *inorganic* matter.

NOTE.—The time required to insure perfect settling after digesting with carbonate of soda solution, varies, with different peats, from 24 hours to several days. With proper care, the results obtained are very satisfactory. Two analyses of No. 6, executed at different times, gave *total insoluble in carbonate of soda*—1st analysis 23.20 *per cent.*; 2d analysis 23.45 *per cent.* These residues yielded respectively 14.30 and 14.15 *per cent.* of ash.

VI. The quantity of *organic matter insoluble in water but soluble in solution of carbonate of soda*, is ascertained by deducting the joint weight of the amounts soluble in water, and insoluble in carbonate of soda, from the total amount of organic matter present. The *inorganic matter insoluble in water, but soluble in carbonate of soda*, is determined by deducting the joint weight of the amounts of inorganic matter soluble in water, and insoluble in carbonate of soda, from the total inorganic matter.

VII. The amount of nitrogen is estimated by the combustion of 1 gramme with soda-lime in an iron tube, collection of the ammonia in a standard solution of sulphuric acid, and determination of the residual free acid by an equivalent solution of caustic potash and a few drops of tincture of cochineal as an indicator.

The results of the analyses are given in the following Tables. Table I. gives the direct results of analysis. In Table II. the analyses are calculated on dry matter, and the nitrogen upon the organic matters. Table III. gives a condensed statement of the external characters and agricultural value* of the samples in their different localities, and the names of the parties supplying them.

* Derived from the communications published in the author's Report. Trans. Conn. State Ag. Soc. 1858 p.p. 101—153.

TABLE I.—COMPOSITION OF CONNECTICUT PEATS AND MUCKS.

From Whom and Whence Received.			Organic Matter: Soluble in water.	Organic Matter: Insol. in water, but soluble in carbonate of soda.	Organic Matter: Insol. in water and carbonate of soda.	Organic Matter: Total.	Inorganic Matter: Soluble in water.	Inorganic Matter: Insol. in water, but soluble in carbonate of soda.	Inorganic Matter: Insol. in water and carbonate of soda.	Inorganic Matter: Total.	Water.	Nitrogen.	Total matters soluble in water.
1. Lewis M. Norton	Goshen	Conn.	17.63		34.79	52.42				35.21	12.37	1.28	1.54
2. " "	"	"	60.02		11.65	71.67				8.00	20.33	1.85	
3. " "	"	"	50.60		29.75	80.35				4.52	15.13	1.90	2.51
5. Messrs. Pond & Miles	Milford	"	65.15		11.95	77.10				3.23	19.67	1.20	1.63
5. " "	"	"	67.75		16.65	84.40				2.00	13.60	.95	3.42
6. Samuel Camp	Plainville	"	43.20		8.90	52.10	14.90		14.30	29.20	18.70	2.10	2.50
7. Russell U. Peck	Berlin	"	38.49		30.51	69.00				13.59	17.41	1.62	2.61
8. Rev. B. F. Northrop	Griswold	"	42.30		10.15	52.45				34.70	12.85	1.31	1.64
9. J. H. Stanwood	Colebrook	"	49.65		7.40	57.05				4.57	38.38	1.23	1.83
10. N. Hart, Jr.	West Cornwall	"	55.11		10.29	65.40				14.89	19.71	2.10	6.20
11. A. L. Loveland	North Granby	"	38.27		2.89	41.16				47.24	11.60	1.00	.75
12. Daniel Buck, Jr.	Poquonnock	"	27.19		48.84	76.03				5.92	18.05	2.40	2.94
13. " "	"	"	33.66		40.51	74.17				8.63	17.20	2.40	1.80
14. Philip Scarborough	Brooklyn	"	51.45		25.00	76.45				7.67	15.88	1.20	1.43
15. Adams White	"	"	54.38		23.14	77.52				9.03	13.45	2.89	5.90
16. Paris Dyer	"	"	18.86		5.02	23.88				67.77	8.35	1.03	2.63
17. Perrin Scarborough	"	"	43.27		16.83	60.10				25.78	14.12	0.86	15.13
18. Geo. K. Virgin	Collinsville	"	2.21	20.57	8.25	31.03	0.32	9.41	48.05	57.78	11.19	0.64	2.53
19. " "	"	"	1.12	9.19	5.10	15.41	0.28	1.08	48.65	50.01	34.58	0.34	1.40
20. " "	"	"	0.72	9.31	3.65	13.68	0.25	0.76	28.20	29.21	57.11	0.28	.97
21. S. Mead	New Haven	"	3.30	40.52	8.20	52.02	2.60	10.02	23.90	36.52	11.46	1.51	5.90
22. Edwin Hoyt	New Canaan	"	2.84	13.42	7.55	23.81	2.72	19.88	46.30	68.90	7.29	0.45	5.56
23. " "	"	"	2·34	13.49	8.05	23.88	1.54	12.42	56.20	70.16	5.96	0.90	3.88
24. " "	"	"	1.15	17.29	8.00	26.44	1.67	14.13	51.10	66.90	6.66	1.01	2.82
25. A. M. Haling	Rockville	"	3.43	52.15	8.65	64.23	0.35	0.16	4.90	5.41	30.36	1.62	3.78
26. " "	"	"	3.87	71.57	8.44	83.88	0.23		1.98	2.21	13.91	1.32	4.10
27. " "	"	"	3.87	44.04	4.25	52.16	0.51	4.07	5.05	9.63	38.21	1.88	4.33
28. Albert Day	Brooklyn	"	2.45	46.25	6.35	55.05	0.32	0.65	5.40	6.37	38.58	0.84	2.77
29. C. Goodyear	New Haven	"	1.80	45.42	10.35	57.57	0.35	7.98	18.80	27.13	15.30	1.68	2.15
30. Rev. Wm. Clift	Stonington	"	3.33	51.68	9.80	64.81	2.82		5.86	8.68	26.51	0.95	6.15
31. Henry Keeler	South Salem	N. Y.	2.13	45.12	12.05	59.30	0.78	3.79	16.70	21.27	19.43	1.57	2.91
32. John Adams	Salisbury	Conn.	1.71	42.87	10.65	55.23	1.02	1.33	14.35	16.70	28.07	1.76	2.73
33. Rev. Wm. Clift	Stonington	"	5.40	16.72	7.25	29.37	7.40	6.40	48.05	61.85	8.78	1.32	2.80
Average			2.06				1.44					1.37	3.72

TABLE II.—COMPOSITION OF CONNECTICUT PEATS AND MUCKS.

Calculated in the dry state: the percentage of nitrogen calculated also on organic matters.

In this table the matters soluble in water and the nitrogen are calculated to two places of decimals; the other ingredients are expressed in round numbers.			ORGANIC MATTER.				INORGANIC MATTER.						
			Soluble in water.	*Insol. in water, but soluble in carbonate of soda.*	*Insol. in water and in carbonate of soda.*	*Total.*	*Soluble in water.*	*Insol. in water, but soluble in carbonate of soda.*	*Insol. in water and in carbonate of soda.*	*Total.*	*Total matters soluble in water.*	*Nitrogen.*	*Nitrogen in per cent. of the organic matter.*
1. Lewis M. Norton	Goshen	Conn.	20		40	60				40	1.75	1.46	2.25
2. " "	"	"	75		15	90				10		2.32	2.58
3. " "	"	"	60		35	95				5	2.95	2.23	2.36
4. Messrs. Pond & Miles	Milford	"	81		15	96				4	2.03	1.49	1.55
5. " "	"	"	79		19	98				2	3.97	1.09	1.12
6. Samuel Camp	Plainville	"	53		11	64	18		18	36	3.08	2.58	4.03
7. Russell U. Peck	Berlin	"	46		37	83				17	3.27	1.96	2.34
8. Rev. B. F. Northrop	Griswold	"	48		11	59				41	1.88	1.50	2.49
9. J. H. Stanwood	Colebrook	"	75		11	86				14	2.77	1.99	2.15
10. N. Hart, Jr	West Cornwall	"	69		13	82				18	7.75	2.61	3.21
11. A. L. Loveland	North Granby	"	43		4	47				53	.85	1.13	2.43
12. Daniel Buck, Jr	Poquonock	"	33		60	93				7	3.58	2.92	3.15
13. " "	"	"	41		49	90				10	2.16	2.89	3.23
14. Philip Scarborough	Brooklyn	"	61		30	91				9	1.70	1.42	1.57
15. Adams White	"	"	63		27	90				10	6.78	3.33	3.72
16. Paris Dyer	"	"	21		5	26				74	2.85	1.12	4.31
17. Perrin Scarborough	"	"	62		8	70				30	17.59	1.00	1.43
18. Geo. K. Virgin	Collinsville	"	2.48	23	9	35	0.35	11	54	65	2.83	0.72	2.06
19. " "	"	"	1.72	14	8	23	.43	2	75	77	2.15	0.51	2.20
20. " "	"	"	1.67	22	8	32	.58	2	66	68	2.25	0.65	2.04
21. Solomon Mead	New Haven	"	3.70	48	9	60	2.92	11	27	40	6.62	1.70	2.90
22. Edwin Hoyt	New Canaan	"	3.05	14	8	26	2.92	21	50	74	6.07	0.48	1.88
23. " "	"	"	2.47	14	8	25	1.63	13	60	75	4.10	0.95	3.76
24. " "	"	"	1.23	18	9	28	1.79	15	55	72	3.02	1.08	3.82
25. A. M. Haling	Rockville	"	4.90	75	12	92	.50		7	8	5.40	2.32	2.52
26. " "	"	"	4.50	83	10	97	.27		2	3	4.77	1.53	1.57
27. " "	"	"	6.24	71	7	84	.82	7	8	16	7.06	3.04	3.64
28. Albert Day	Brooklyn	"	4.01	76	10	90	.52	1	8	10	4.53	1.36	1.52
29. C. Goodyear	New Haven	"	2.11	54	12	68	.40	9	22	32	2.51	1.98	2.91
30. Rev. Wm. Clift	Stonington	"	4.56	71	13	88	3.86		8	12	8.42	1.29	1.46
31. Henry Keeler	South Salem	N. Y.	2.66	56	15	73	.97	5	21	27	3.63	1.98	2.64
32. John Adams	Salisbury	Conn.	2.37	59	15	76	1.40	2	20	24	3.77	2.44	3.18
33. Rev. Wm. Clift	Stonington	"	5.93	18	8	32	8.13	7	53	68	11.06	1.44	4.49

TABLE III.—DESCRIPTION, ETC., OF PEATS AND MUCKS.

No.		*Color.*		*Condition at Time of Analysis, Reputed Value, etc.*
1.	Lewis M. Norton	chocolate-brown,	air-dry,	tough, compact, heavy; from bottom; 3 to 4 feet deep; very good in compost.
2.	" "	" "	"	" " heavier than 1, from near surface; " "
3.	" "	light-brown,	"	coherent but light, from between 1 and 2, " "
4.	Messrs. Pond & Miles	chocolate-brown,	"	" " surface peat, considered better than No. 5; good in compost.
5.	" "	brownish-red,	"	very light and loose in texture, from depth of 3 feet, " "
6.	Samuel Camp	black,	"	hard lumps, half as good as yard manure, in compost equal to yard manure.
7.	Russell U. Peck	chocolate-brown,	"	is good fresh, long exposed, half as good as barn-yard manure.
8.	Rev. B. F. Northrop	grayish-brown,	"	light, easily crushed masses containing sand, has not been used alone, good in compost.
9.	J. H. Stanwood	chocolate-brown,	moist,	hard lumps, used fresh good after first year: excellent in compost.
10.	N. Hart, Jr	brownish-black,	air-dry,	" excellent in compost.
11.	A. L. Loveland	black,	"	" contains grains of coarse sand.
12.	Daniel Buck, Jr	chocolate-brown,	"	coherent cakes, good as top-dressing on grass when fresh; excellent in compost.
13.	" "	" "	"	light surface layers of No. 12.
14.	Phillip Scarborough		"	after exposure over winter, has one-third value of yard-manure.
15.	Adams White	chocolate-brown,	"	hard lumps, good in compost, causes great growth of straw.
16.	Paris Dyer	grayish-black,	"	easily crushed lumps, largely admixed with soil.
17.	Perrin Scarborough	chocolate-brown,	"	well-characterized "vitriol peat;" in compost, after 1 year's exposure, gives indifferent results.
18.	Geo. K. Virgin	light brownish-gray,	"	light, coherent, surface peat; sample long exposed; astonishing results on sandy soil.
19.	" "	chocolate-brown,	moist,	crumbly, contains much sand, four feet from surface.
20.	" "	black,	wet.	
21.	Solomon Mead	grayish-brown,	air-dry,	light, porous, coherent from grass roots; long weathered, good; fresh, better in compost.
22.	Edwin Hoyt	brownish-gray,	"	loose, light, much mixed with soil, good in compost.
23.	" "	"	"	No. 22 saturated with horse urine, darker than No. 22.
24.	" "	"	"	No. 22 composted with white fish, darker than No. 23; fish-bones evident.
25.	A. M. Haling	chocolate-brown,	moist,	fresh dug.
26.	" "	" "	air-dry,	No. 25 after two year's weathering.
27.	" "	" "	moist,	fresh dug, good substitute for yard manure as top-dressing on grass.
28.	Albert Day	dark-brown,	"	coherent and hard; fresh dug, but from surface where weathered; injurious to crops; vitriol peat. (?)
29.	C. Goodyear	black,	air-dry,	very hard tough cakes; when fresh dug, "as good as cow dung."
30.	Rev. Wm. Clift	chocolate-brown,	moist,	from an originally fresh water bog, broken into 100 years ago by tide, now salt-marsh; good after weathering.
31.	Henry Keeler	light-brown,	air-dry,	leaf-muck, friable; when fresh, appears equal to good yard manure.
32.	John Adams	"	moist,	overlies shell marl, fresh or weathered does not compare with ordinary manure.
33.	Rev. Wm. Clift	dark ash-gray,	air-dry,	from bottom of salt ditch, where tide flows daily; contains sulphate of iron.

PART III.

ON PEAT AS FUEL.

1.—*Kinds of peat that make the best fuel.*

The value of peat for fuel varies greatly, like its other qualities. Only those kinds which can be cut out in the shape of coherent blocks, or which admit of being artificially formed into firm masses, are of use in ordinary stoves and furnaces. The powdery or friable surface peat, which has been disintegrated by frost and exposure, is ordinarily useless as fuel, unless it be rendered coherent by some mode of preparation. Unripe peat which contains much undecomposed moss or grass roots, which is therefore very light and porous, is in general too bulky to make an effective heating material before subjection to mechanical treatment.

The best peat for burning, is that which is most free from visible fiber or undecomposed vegetable matters, which has therefore a homogeneous brown or black aspect, and which is likewise free from admixture of earthy substances in the form of sand or clay. Such peat is unctuous when moist, shrinks greatly on drying, and forms hard and heavy masses when dry. It is usually found at a considerable depth, where it has been subjected to pressure, and then has such consistence as to admit of cutting out in blocks; or it may exist as a black mud or paste at the bottom of bogs and sluices.

The value of peat as fuel stands in direct ratio to its content of carbon. We have seen that this ranges from 51 to 63 *per cent. of the organic matter*, and the increase of carbon is related to its ripeness and density. The poorest, youngest peat, has the same proportion of carbon as exists in wood. It does not, however, follow that its heating power is the same. The various kinds of wood have essentially the same proportion of carbon, but their heating power is very different. The close textured woods—those which weigh the most per cord—make the best fuel for most purposes. We know that a cord of hickory will produce twice as much heat as a cord of bass-wood. Peat, though having the same or a greater proportion of carbon, is generally inferior to wood on account of its occupying a greater bulk for a given weight, a necessary result of its porosity. The best qualities of peat, or poor kinds artificially condensed, may, on the other hand, equal or exceed wood in heating power, bulk for bulk. One reason that peat is, in general, inferior to wood in heating effect, lies in its greater content of incombustible ash. Wood has but 0.5 to 1.5 *per cent.* of mineral matters, while peat contains usually 5 to 10 *per cent.*, and often more. The oldest, ripest peats are those which contain the most carbon, and have at the same time the greatest compactness. From these two circumstances they make the best fuel.

It thus appears that peat which is light, loose in structure, and much mixed with clay or sand, is a poor or very poor article for producing heat: while a dense pure peat is very good.

A great drawback to the usefulness of most kinds of peat-fuel, lies in their great friability. This property renders them unable to endure transportation. The blocks of peat which are commonly used in most parts of Germany as fuel, break and crumble in handling, so that they cannot be carried far without great waste. Besides, when put into a stove, there can only go on a slow smouldering combustion as would happen in cut tobacco or saw-dust. A free-burning fuel must exist in compact lumps or blocks, which so retain their form and solidity, as to admit of a rapid draught of air through the burning mass.

The bulkiness of ordinary peat fuel, as compared with hard wood, and especially with coal, likewise renders transportation costly, especially by water, where freights are charged by bulk and not by weight, and renders storage an item of great expense.

The chief value of that peat fuel, which is simply cut from the bog, and dried without artificial condensation, must be for the domestic use of the farmer or villager who owns a supply of it not far from his dwelling, and can employ his own time in getting it out. Though worth perhaps much less cord for cord when dry than hard wood, it may be cheaper for home consumption than fuel brought from a distance.

Various processes have been devised for preparing peat, with a view to bringing it into a condition of density and toughness, sufficient to obviate its usual faults, and make it compare with wood or even with coal in heating power.

The efforts in this direction have met with abundant success as regards producing a good fuel. In many cases, however, the cost of preparation has been too great to warrant the general adoption of these processes. We shall recur to this subject on a subsequent page, and give an account of the methods that have been proposed or employed for the manufacture of condensed peat fuel.

2.—*Density of Peat.*

The apparent* specific gravity of peat in the air-dry state, ranges from 0.11 to 1.03. In other words, a full cubic foot weighs from one-tenth as much as, to slightly more than a cubic foot of water, = 62⅓ lbs. Peat, which has a specific gravity of but 0.25, may be and is employed as fuel. A full cubic foot of it will weigh about 16 lbs. In Germany, the cubic foot of "good ordinary peat" in blocks,† ranges from 15 to 25 lbs. in weight, and is employed for domestic purposes. The heavier peat, weighing 30 or more lbs. per cubic foot in blocks, is used for manufacturing and metallurgical purposes, and for firing locomotives.

Karmarsch has carefully investigated more than 100 peats belonging to the kingdom of Hanover, with reference to their heating effect. He classifies them as follows:—

A. *Turfy peat*, (*Rasentorf*,) consisting of slightly decomposed mosses and other peat-producing plants, having a yellow or yellowish-brown color, very soft, spongy and

* The apparent specific gravity here means the weight of the mass, —the air-filled cavities and pores included — as compared with an equal bulk of water. The real specific gravity of the *peat itself* is always greater than that of water, and all kinds of peat will sink in water when they soak long enough, or are otherwise treated so that all air is removed.

† The "full" cubic foot implies a cubic foot having no cavities or waste space, such as exist in a pile, made up of numerous blocks. If a number of peat blocks be put into a box and shaken together, the empty space between the more or less irregular blocks, may amount to 46 *per cent.* of the whole; and when closely packed, the cavities amount to 30 *per cent.*, according to the observations of *Wasserzieher*. (*Dingler's Journal*, Oct., 1864, p. 118.) Some confusion exists in the statements of writers in regard to this matter, and want of attention to it, has led to grave errors in estimating the weight of fuel.

elastic, sp. gr. 0.11 to 0.26, the full English cubic foot weighing from 7 to 16 lbs.

B. *Fibrous peat*, unripe peat, which is brown or black in color, less elastic than turfy peat, the fibres either of moss, grass, roots, leaves, or wood, distinguishable by the eye, but brittle, and easily broken; sp. gr. 0.24 to 0.67, the weight of a full cubic foot being from 15 to 42 lbs.

C. *Earthy peat.*—Nearly or altogether destitute of fibrous structure, drying to earth-like masses which break with more or less difficulty, giving lustreless surfaces of fracture; sp. gr. 0.41 to 0.90, the full cubic foot weighing, accordingly, from 25 to 56 lbs.

D. *Pitchy peat*, (*Pechtorf*,) dense; when dry, hard; often resisting the blows of a hammer, breaking with a smooth, sometimes lustrous fracture, into sharp-angled pieces. Sp. gr. 0.62 to 1.03, the full cubic foot weighing from 38 to 55 lbs.

In Kane and Sullivan's examination of 27 kinds of Irish peat, the specific gravities ranged from 0.274 to 1.058.

3.—*Heating power of peat as compared with wood and anthracite.*

Karmarsch found that in absolute heating effect

100 lbs. of	turfy, air-dry peat, on the average			=	95 lbs.	of	pine wood.
"	fibrous	"	"	"	=	108	"	"
"	earthy	"	"	"	=	104	"	"
"	pitchy	"	"	"	=	111	"	"

The comparison of heating power by bulk, instead of weight, is as follows: —

100 cubic	ft. of	turfy peat, on the average*			=	33	cubic	ft. of	pine wood,	in sticks.
"	"	fibrous	"	"	=	90	"	"	"	"
"	"	earthy	"	"	=	145	"	"	"	"
"	"	pitchy	"	"	=	184	"	"	"	"

* The *waste space* in peat and wood as commonly piled, is probably included here in the statement, and is usually about the same in both; viz.: not far from 40 *per cent.*

According to Brix, the weight per English cord and relative heating effect of several air-dry peats—the heating power of an equal bulk of oak wood being taken at 100 as a standard—are as follows, *bulk for bulk:* *

	Weight per cord.	*Heating effect.*
Oak wood	4150 lbs.	100
Peat from Linum, 1st quality, dense and pitchy	3400 "	70
" " 2d " fibrous	2900 "	55
" " 3d " turfy	2270 "	53
Peat from Buechsenfeld, 1st quality, pitchy, very hard and heavy,	3400 lbs.	74
" " 2d "	2730 "	64

These statements agree in showing, that, while weight for weight, the ordinary qualities of peat do not differ much from wood in heating power; the heating effect of *equal bulks* of this fuel, as found in commerce, may vary extremely, ranging from one-half to three quarters that of oak wood.

Condensed peat may be prepared by machinery, which will weigh more than hard wood, bulk for bulk, and whose heating power will therefore exceed that of wood.

Gysser gives the following comparisons of a good peat with various German woods and charcoals, equal weights being employed, and split beech wood, air-dry, assumed as the standard.†

Beech wood, split, air dry	1.00
Peat, condensed by Weber's & Gysser's method,‡ air-dried, with 25 *per cent.* moisture	1.00
Peat, condensed by Weber's & Gysser's method, hot-dried, with 10 *per cent.* moisture	1.48
Peat-charcoal, from condensed peat	1.73
The same peat, simply cut and air-dried	0.80
Beech-charcoal	1.90
Summer-oak wood	1.18
Birch wood	0.95
White pine wood	0.72
Alder	0.65
Linden	0.65
Red pine	0.61
Poplar	0.50

* See note on the preceding page.

† *Der Torf, etc.*, *S.* 43. ‡ See page 00.

Some experiments have been made in this country on the value of peat as fuel. One was tried on the N. Y. Central Railroad, Jan. 3, 1866. A locomotive with 25 empty freight cars attached, was propelled from Syracuse westward—the day being cold and the wind ahead—at the rate of 16 miles the hour. The engineer reported that "the peat gave us as much steam as wood, and burnt a beautiful fire." The peat, we infer, was cut and prepared near Syracuse, N. Y.

In one of the pumping houses of the Nassau Water Department of the City of Brooklyn, an experiment has been made for the purpose of comparing peat with anthracite, for the results of which I am indebted to the courtesy of Moses Lane, Esq., Chief Engineer of the Department.

Fire was started under a steam boiler with wood. When steam was up, the peat was burned—its quantity being 1743 lbs., or 18 barrels—and after it was consumed, the firing was continued with coal. The pressure of steam was kept as nearly uniform as possible throughout the trial, and it was found that with 1743 lbs. of peat the engine made 2735 revolutions, while with 1100 lbs. of coal it made 3866 revolutions. In other words, 100 lbs. of coal produced $351\frac{45}{100}$ revolutions, and 100 lbs. of peat produced $156\frac{91}{100}$ revolutions. One pound of coal therefore equalled $2\frac{24}{100}$ lbs. of peat in heating effect. The peat burned well and generated steam freely.

Mr. Lane could not designate the quality of the peat, not having been able to witness the experiment.

These trials have not, indeed, all the precision needful to fix with accuracy the comparative heating effect of the fuels employed; for a furnace, that is adapted for wood, is not necessarily suited to peat, and a coal grate must have a construction unlike that which is proper for a peat fire; nevertheless they exhibit the relative merits of wood, peat, and anthracite, with sufficient closeness for most practical purposes.

Two considerations would prevent the use of ordinary cut peat in large works, even could two and one-fourth tons of it be afforded at the same price as one ton of coal. The Nassau Water Department consumes 20,000 tons of coal yearly, the handling of which is a large expense, six firemen being employed to feed the furnaces. To generate the same amount of steam with peat of the quality experimented with, would require the force of firemen to be considerably increased. Again, it would be necessary to lay in, under cover, a large stock of fuel during the summer, for use in winter, when peat cannot be raised. Since a barrel of this peat weighed less than 100 lbs., the short ton would occupy the volume of 20 barrels; as is well known, a ton of anthracite can be put into 8 barrels. A given weight of peat therefore requires 2½ times as much storage room, as the same weight of coal. As 2¼ tons of peat, in the case we are considering, are equivalent to but one ton of coal in heating effect, the winter's supply of peat fuel would occupy 5⅝ times the bulk of the same supply in coal, admitting that the unoccupied or air-space in a pile of peat is the same as in a heap of coal. In fact, the calculation would really turn out still more to the disadvantage of peat, because the air-space in a bin of peat is greater than in one of coal, and coal can be excavated for at least two months more of the year than peat.

It is asserted by some, that, because peat can be condensed so as to approach anthracite in specific gravity, it must, in the same ratio, approach the latter in heating power. Its effective heating power is, indeed, considerably augmented by condensation, but no mechanical treatment can increase its percentage of carbon or otherwise alter its chemical composition; hence it must forever remain inferior to anthracite.

The composition and density of the best condensed peat is compared with that of hard wood and anthracite in the following statement:—

In 100 parts.	*Carbon.*	*Hydrogen.*	*Oxygen and Nitrogen.*	*Ash.*	*Water.*	*Specific Gravity.*
Wood	39.6	4.8	34.8	0.8	20.0	0.75
Condensed peat	47.2	4.9	22.9	5.0	20.0	1.20
Anthracite	91.3	2.9	2.8	3.0		1.40

In combustion in ordinary fires, the *water* of the fuel is a source of waste, since it consumes heat in acquiring the state of vapor. This is well seen in the comparison of the same kind of peat in different states of dryness. Thus, in the table of Gysser, (page 97) Weber's condensed peat, containing 10 *per cent.* of moisture, surpasses in heating effect that containing 25 *per cent.* of moisture, by nearly one-half.

The *oxygen* is a source of waste, for heat as developed from fuel, is chiefly a result of the chemical union of atmospheric or free oxygen, with the carbon and hydrogen of the combustible. The oxygen of the fuel, being already combined with carbon and hydrogen, not only cannot itself contribute to the generation of heat, but neutralizes the heating effect of those portions of the carbon and hydrogen of the fuel with which it remains in combination. The quantity of heating effect thus destroyed, cannot, however, be calculated with certainty, because physical changes, viz: the conversion of solids into gases, not to speak of secondary chemical transformations, whose influence cannot be estimated, enter into the computation.

Nitrogen and ash are practically indifferent in the burning process, and simply impair the heating value of fuel in as far as they occupy space in it, and make a portion of its weight, to the exclusion of combustible matter.

Again, as regards density, peat is, in general, considerably inferior to anthracite. The best uncondensed peat has a specific gravity of 0.90. Condensed peat usually does not exceed 1.1. Sometimes it is made of sp. gr. 1.3. Assertions to the effect of its acquiring a density of 1.8, can hardly be credited of pure peat, though a considerable admixture of sand or clay might give such a result.

The comparative heating power of fuels is ascertained by burning them in an apparatus, so constructed, that the heat generated shall expend itself in evaporating or raising the temperature of a known quantity of water.

*The amount of heat that will raise the temperature of one gramme of water, one degree of the centigrade thermometer, is agreed upon as the unit of heat.**

In the complete combustion of carbon in the form of charcoal or gas-coal, there are developed 8060 units of heat. In the combustion of one gramme of hydrogen gas, 34,210 units of heat are generated. The heating effect of hydrogen is therefore 4.2 times greater than that of carbon. It was long supposed that the heating effect of compound combustibles could be calculated from their elementary composition. This view is proved to be erroneous, and direct experiment is the only satisfactory means of getting at the truth in this respect.

The data of Karmarsch, Brix, and Gysser, already given, were obtained by the experimental method. They were, however, made mostly on a small scale, and, in some cases, without due regard to the peculiar requirements of the different kinds of fuel, as regards fire space, draught, etc. They can only be regarded as approximations to the truth, and have simply a comparative value, which is, however, sufficient for ordinary purposes.

* On account of the great convenience of the decimal weights and measures, and their nearly universal recognition by scientific men, we have adopted them here. The gramme = 15 grains ; 5 degrees centigrade = 9 degrees Fahrenheit.

The general results of the investigations hitherto made on all the common kinds of fuel, are given in the subjoined statement. The comparison is made in units of heat, and refers to equal weights of the materials experimented with.

HEATING POWER OF DIFFERENT KINDS OF FUEL.

Air-dry Wood	2800	
" Peat	2500	3000
Perfectly dry Wood	3600	
" " Peat	3000	4000
Air-dry Lignite or Brown Coal	3300	4200
Perfectly dry Lignite or Brown Coal	4000	5000
Bituminous Coal	3800	7000
Anthracite	7500	
Wood Charcoal	6300	7500
Coke	6500	7600

4.—*Modes of Burning Peat.*

In the employment of peat fuel, regard must be had to its shape and bulk. Commonly, peat is cut or moulded into blocks or sods like bricks, which have a length of 8 to 18 inches; a breadth of 4 to 6 inches, and a thickness of 1½ to 3 inches. Machine peat is sometimes formed into circular disks of 2 to 3 inches diameter, and 1 to 2 inches thickness and thereabouts. It is made also in the shape of balls of 2 to 3 inches diameter. Another form is that of thick-walled pipes, 2 to 3 inches in diameter, a foot or more long, and with a bore of one-half inch.

Flat blocks are apt to lie closely together in the fire, and obstruct the draft. A fire-place, constructed properly for burning them, should be shallow, not admitting of more than two or three layers being superposed. According to the bulkiness of the peat, the fire-place should be roomy, as regards length and breadth.

Fibrous and easily crumbling peat is usually burned upon a hearth, *i. e.* without a grate, either in stoves or open fire-places. Dense peat burns best upon a grate, the bars of which should be thin and near together, so that the air have access to every part of the fuel. The denser and tougher the peat, and the more its shape corresponds with that usual to coal, the better is it adapted for use in our ordinary coal stoves and furnaces.

5.—*Burning of broken peat.*

Broken peat—the fragments and waste of the cut or moulded blocks, and peat as obtained by plowing and harrowing the surface of drained peat-beds—may be used to advantage in the ***stair grate***, fig. 1, which was introduced

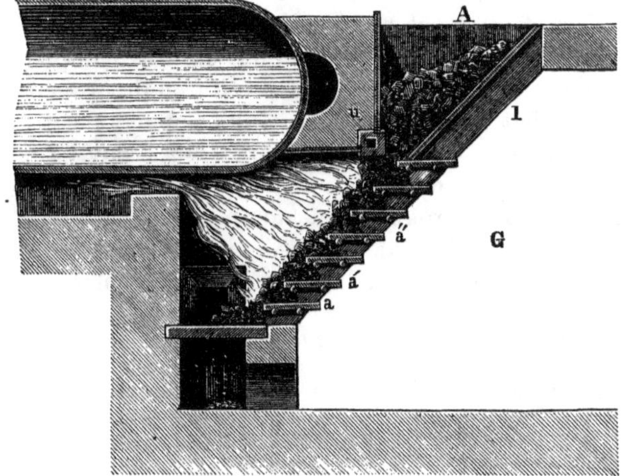

Fig. 1.—STAIR GRATE.

some years ago in Austria, and is adapted exclusively for burning finely divided fuel. It consists of a series of thin iron bars 3 to 4 inches wide, *a*, *a*, *a*,.... which are arranged above each other like steps, as shown in the figure. They are usually half as long as the grate is wide, and are supported at each end by two side pieces or walls, *l*. Below, the grate is closed by a heavy iron plate. The fuel is placed in the hopper *A*, which is kept filled, and from which it falls down the incline as rapidly as it is consumed. The air enters from the space *G*, and is regulated by doors, not shown in the cut, which open into it. The masonry is supported at *u*, by a hollow iron beam. Below, a lateral opening serves for clearing out the ashes. The effect of the fire depends upon the width of the throat of the hopper at *u*, which regulates the supply of fuel to the grate, and upon the inclination of the latter. The throat is usually from 6 to 8 inches wide, according to the nature of the fuel. The inclination of the grate is 40 to 45° and, in general, should be that which is assumed by the sides of a pile of the fuel to be burned, when it is thrown up into a heap. This grate ensures complete combustion of fuel that would fall through ordinary grates, and that would merely smoulder upon a hearth. The fire admits of easy regulation, the ashes may be removed and the fuel may be supplied without *checking the fire.* Not only broken peat, but coal dust, saw dust, wood turnings and the like may be burned on this grate. The figure represents it as adapted to a steam boiler.

6.—*Hygroscopic water of peat fuel.*

The quantity of water retained by air-dried peat appears to be the same as exists in air-dried wood, viz., about 20 *per cent.* The proportion will vary however according to the time of seasoning. In thoroughly seasoned wood or peat, it may be but 15 *per cent.*; while in the poorly dried material it may amount to 25 or more *per cent.* When *hot-dried*, the proportion of water may be reduced to 10 *per cent.*, or less.

When peat is still moist, it gathers water rapidly from damp air, and in this condition has been known to burst the sheds in which it was stored, but after becoming dry to the eye and feel, it is but little affected by dampness, no more so, it appears, than seasoned wood.

7.—*Shrinkage.*

In estimating the value and cost of peat fuel, it must be remembered that peat shrinks greatly in drying, so that three to five cords of fresh peat yield but one cord of dry peat. When the fiber of the peat is broken by the hand, or by machinery, the shrinkage is often much greater, and may sometimes amount to seven-eighths of the original volume.—*Dingler's Journal, Oct.* 1864, *S.* 68.

The difference in weight between fresh and dry peat is even greater. Fibrous peat, fresh from the bog, may contain ninety *per cent.* of water, of which seventy *per cent.* must evaporate before it can be called dry. The proportion of water in earthy or pitchy peat is indeed less; but the quantity is always large, so that from five to nine hundred weight of fresh peat must be lifted in order to make one hundred weight of dry fuel.

8.—*Time of excavation, and drying.*

Peat which is intended to be used after simply drying, must be excavated so early in the season that it shall become dry before frosty weather arrives: because, if frozen when wet, its coherence is destroyed, and on thawing it falls to a powder useless for fuel.

Peat must be dried with certain precautions. If a block of fresh peat be exposed to hot sunshine, it dries and shrinks on the surface much more rapidly than within: as a consequence it cracks, loses its coherence, and the block is easily broken, or of itself falls to pieces. In Europe, it is indeed customary to dry peat without shelter, the loss by too rapid drying not being greater than the expense of building and maintaining drying sheds. There however the sun is not as intense, nor the air nearly so dry, as it is here. Even there, the occurrence of an unusually hot summer, causes great loss. In our climate, some shelter would be commonly essential unless the peat be dug early in the spring, so as to lose the larger share of its water before the hot weather; or, as would be best of all, in the autumn late enough to escape the heat, but early enough to ensure such dryness as would prevent damage by frost. The peculiarities of climate must decide the time of excavating and the question of shelter.

The point in drying peat is to make it lose its water gradually and regularly, so that the inside of each block shall dry nearly as fast as the outside.

Some of the methods of hot-drying peat, will be subsequently noticed.

Summer or fall digging would be always advantageous on account of the swamps being then most free from water. In Bavaria, peat is dug mostly in July and the first half of August.

9.—*Drainage.*

When it is intended to raise peat fuel *in the form of blocks*, the bog should be drained no more rapidly than it is excavated. Peat, which is to be worth cutting in the spring, must be covered with water during the winter, else it is pulverized by the frost. So, too, it must be protected against drying away and losing its coherency in summer, by being kept sufficiently impregnated with water.

In case an extensive bog is to be drained to facilitate the cutting out of the peat for use as fuel, the canals that carry off the water from the parts which are excavating, should be so constructed, that on the approach of cold weather, the remaining peat may be flooded again to the usual height.

In most of the smaller swamps, systematic draining is unnecessary, the water drying away in summer enough to admit of easy working.

In some methods of preparing or condensing peat by machinery, it is best or even needful to drain and air-dry the peat, preliminary to working. By draining, the peat settles, especially on the borders of the ditches, several inches, or even feet, according to its nature and depth. It thus becomes capable of bearing teams and machinery, and its density is very considerably augmented.

10.—*The Cutting of Peat.—a. Preparations.*

In preparing to raise peat fuel from the bog, the surface material, which from the action of frost and sun has been pulverized to "muck," or which otherwise is full of roots and undecomposed matters, must be removed usually to the depth of 12 to 18 inches. It is only those portions of the peat which have never frozen nor become dry, and are free from coarse fibers of recent vegetation, that can be cut for fuel.

Peat fuel must be brought into the form of blocks or masses of such size and shape as to adapt them to use in our common stoves and furnaces. Commonly, the peat is of such consistence in its native bed, that it may be cut out with a spade or appropriate tool into blocks having more or less coherence. Sometimes it is needful to take away the surplus water from the bog, and allow the peat to settle and drain a while before it can be cut to advantage.

When a bog is to be opened, a deep ditch is run from an outlet or lowest point a short distance into the peat bed, and the working goes on from the banks of this ditch. It is important that system be followed in raising the peat, or there will be great waste of fuel and of labor.

If, as often happens, the peat is so soft in the wet season as to break on the vertical walls of a ditch and fill it, at the same time dislocating the mass and spoiling it for cutting, it is best to carry down the ditch in terraces, making it wide above and narrow at the bottom.

b. Cutting by hand.

The simplest mode of procedure, consists in laying off a "field" or plot of, say 20 feet square, and making vertical cuts with a sharp spade three or four inches deep from end to end in parallel lines, as far apart as it is proposed to make the breadth of the peats or sods, usually four to five inches. Then, the field is cut in a similar manner in lines at right angles to the first, and at a distance that shall be the length of the peats, say 18 to 20 inches. Finally, the workman lifts the peats by horizontal thrusts of his spade, made at a depth of three inches. The sods as lifted, are placed on a light barrow or upon a board or rack, and are carried off to a drying ground, near at hand, where they are laid down flatwise to drain and dry. In Ireland, it is the custom, after the peats have lain thus for a fortnight or so, to "foot" them, i. e. to place them on end close together; after further drying the "footing" is succeeded by "clamping," which is building the sods up into stacks of about twelve to fifteen feet long, four feet wide at bottom, narrowing to one foot at top, with a height of four to five feet. The outer turfs are inclined so as to shed the rain. The peat often remains in these clamps on the bog until wanted for use, though in rainy seasons the loss by crumbling is considerable.

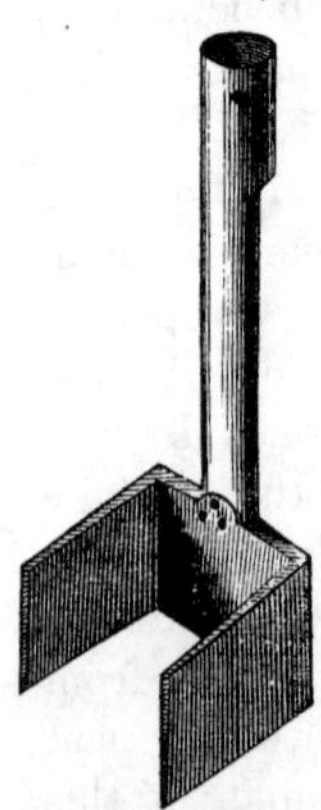

Fig. 2.—GERMAN PEAT-KNIFE.

Other modes of lifting peat, require tools of particular construction.... In Germany it is common to excavate by *vertical* thrusts of the tool, the cutting part of which is represented above, fig. 2. This tool is pressed down into the peat to a depth corresponding to the thickness of the required block: its three edges cut as many sides of the block, and the bottom is then broken or torn out by a prying motion.

In other cases, this or a similar tool is forced down by help of the foot as deeply into the peat as possible by a workman standing above, while a second man in the ditch cuts out the blocks of proper thickness by means of a sharp spade thrust horizontally. When the peats are taken out to the depth of the first vertical cutting, the knife is used again from above, and the process is thus continued as before, until the bottom of the peat or the desired depth is reached.

In Ireland, is employed the "slane," a common form of which is shown in fig. 3, it being a long, narrow and sharp spade, 20 inches by six, with a wing at right angles to the blade.

Fig. 3.—IRISH SLANE.

The peats are cut by one thrust of this instrument which is worked by the arms alone. After a vertical cut is made by a spade, in a line at right angles to a bank of peat, the slane cuts the bottom and other side of the block; while at the end the latter is simply lifted or broken away.

Peat is most easily cut in a vertical direction, but when, as often happens, it is made up of layers, the sods are likely to break apart where these join. Horizontal cutting is therefore best for stratified peat.

System employed in East Friesland.—In raising peat, great waste both of labor and of fuel may easily occur as the result of random and unsystematic methods of working. For this reason, the mode of cutting peat, followed in the extensive moors of East Friesland, is worthy of particular description. There, the business is pursued systematically on a plan, which, it is claimed, long experience * has developed to such perfection that the utmost economy of time and labor is attained. The cost of producing marketable peat in East Friesland in 1860, was one silver groschen=about 2½ cents, per hundred weight; while at that time, in Bavaria, the hundred weight cost three times as much when fit for market; and this, notwithstanding living and labor are much cheaper in the latter country.

The method to be described, presupposes that the workmen are not hindered by water, which, in most cases, can be easily removed from the high-moors of the region. The peat is worked in long stretches of 10 feet in width, and 100 to 1000 paces in length: each stretch or plot is excavated at once to a considerable depth and to its full width. Each successive year the excavation is widened by 10 feet, its length remaining the same. Sometimes, unusual demand leads to more rapid working; but the width of 10 feet is adhered to for each cutting, and, on account of the labor of carrying the peats, it is preferred to extend the length rather than the width.

Assuming that the peat bed has been opened by a previous cutting, to the depth of 5½ feet, and the surface muck and light peat, 1½ feet thick, have been thrown into the excavation of the year before—a new plot is worked by five men as follows.

---

* Pliny, Hist. Nat. (Lib. XVI. 1) expresses his pity for the "miserable people" living in East Friesland and vicinity in his day, who "dug out with the hands a moor earth, which, dried more by wind than sun, they used for preparing their food and warming their bodies:" *captum manibus lutum ventis magis quam sole siccantis, terra cibos et rigentia septembrione viscera sua urunt.*

As regards the "*misera gens*," it should be said that rich grain fields and numerous flourishing villages have occupied for several centuries large portions of the Duevel moor near Bremen.

One man, the "Bunker," removes from the surface, about two inches of peat, disintegrated by the winter's frost, throwing it into last year's ditch.

Following him, come two "Diggers," of whom one stands on the surface of the peat, and with a heavy, long handled tool, cuts out the sides and end of the blocks, which are about seventeen by five inches; while the other stands in the ditch, and by horizontal thrusts of a light, sharp spade, removes the sods, each of five and a half inches thickness, and places them on a small board near by. Each block of peat has the dimensions of one fourth of a cubic foot, and weighs about 13 pounds. Two good workmen will raise 25 such peats, or 6¼ cubic feet, per minute.

A fourth man, the "Loader," puts the sods upon a wheel-barrow, always two rows of six each, one upon the other, and—

A fifth, the "Wheeler," removes the load to the drying ground, and with some help from the Bunker, disposes them flatwise in rows of 16 sods wide, which run at right angles to the ditch, and, beginning at a little more than 10 feet from the latter, extend 50 feet.

The space of 10 feet between the plot that is excavating, and the drying ground, is, at the same time, cleared of the useless surface muck by the Bunker, in preparation for the next year's work.

With moderate activity, the five men will lift and lay out 12,000 sods (3000 cubic feet,) daily, and it is not uncommon that five first-rate hands get out 16,800 peats (4200 cubic feet,) in this time.

A gang of five men, working as described, suffices for cutting out a bed of four feet of solid peat. When the excavation is to be made deeper, a sixth man, the "Hanker," is needful for economical work; and with his help the cutting may be extended down to nine and a half feet; i. e.

through eight feet of solid peat. The cutting is carried down at first, four feet as before, but the peats are carried 50 feet further, in order to leave room for those to be subsequently lifted. The "Hanker" aids here, with a second wheel-barrow. In taking out the lower peat, the "Hanker" stands on the bottom of the first excavation, receives the blocks from the Diggers, on a broad wooden shovel, and hands them up to the Loader; while the Wheeler, having only the usual distance to carry them, lays them out in the drying rows without difficulty.

After a little drying in the rows, the peats are gradually built up into narrow piles, like a brick wall of one and a half bricks thickness. These piles are usually raised by women. They are made in the spaces between the rows, and are laid up one course at a time, so that each block may dry considerably, before it is covered by another. A woman can lay up 12,000 peats daily — the number lifted by 5 men—and as it requires about a month of good weather to give each course time (two days) to dry, she is able to pile for 30 gangs of workmen. If the weather be very favorable, the peats may be stacked or put into sheds, in a few days after the piling is finished. Stacking is usually practised. The stacks are carefully laid up in cylindrical form, and contain 200 to 500 cubic feet. When the stacks are properly built, the peat suffers but little from the weather.

According to Schrœder, from whose account (Dingler's Polytechnisches Journal, Bd. 156, S.128) the above statements are derived, the peats excavated under his direction, in drying thoroughly, shrank to about one-fourth of their original bulk (became 12 inches × 3 inches × 3 inches,) and to one-seventh or one-eighth of their original weight.

c. Machines for Cutting Peat.

In North Prussia, the Peat Cutting Machine of Brosowsky, see fig. 4, is extensively employed. It consists of a cutter, made like the four sides of a box, but with oblique edges, *a*, which by its own weight, and by means of a crank and rack-work, operated by men, is forced down into the peat to a depth that may reach 20 feet. It can cut only at the edge of a ditch or excavation, and when it has penetrated sufficiently, a spade like blade, *d*, is driven under the cutter by means of levers *c*, and thus a mass is loosened, having a vertical length of 10 feet or more, and whose other dimensions are about 24 × 28 inches. This is lifted by reversing the crank motion, and is then cut up by the spade into blocks of 14 inches × 6 inches × 5 inches. Each parallelopipedon of peat, cut to a depth of 10 feet, makes 144 sods, and this number can be cut in less than 10 minutes. Four hands will cut and lay out to dry, 12,000 to 14,000 peats daily, or 3100 cubic feet. One great advantage of this machine consists in the circumstance that it can be used to raise peat from below the surface of water, rendering drainage in many cases unnecessary. Independently of this, it appears to be highly labor saving, since 1300 machines were put to use in Mecklenburg and Pomerania in about 5 years from its introduction. The Mecklenburg moors are now traversed by canals, cut by this machine, which are used for the transportation of the peat to market.*

Lepreux in Paris, has invented a similar but more complicated machine, which is said to be very effective in its operation. According to Hervé Mangon, this machine, when worked by two men, raises and cuts 40,000 peats daily, of which seven make one cubic foot, equal to 5600

* For further account and plans of this machine see Dingler's Polytechnisches Journal, Bd. 176, S. 336.

(114) Fig. 4.—BROSOWSKY'S PEAT CUTTER.

cubic feet. The saving in expense by using this machine* is said to be 70 *per cent.*, when the peat to be raised is under water.

11.—*The Dredging of Peat.*

When peat exists, not as a coherent more or less fibrous mass, but as a paste or mud, saturated with water, it cannot be raised and formed by the methods above described. In such cases the peat is dredged from the bottom of the bog by means of an iron scoop, like a pail with sharp upper edges, which is fastened to a long handle. The bottom is made of coarse sacking, so that the water may run off. Sometimes, a stout ring of iron with a bag attached, is employed in the same way. The fine peat is emptied from the dredge upon the ground, where it remains, until the water has been absorbed or has evaporated, so far as to leave the mass somewhat firm and plastic. In the mean time, a drying bed is prepared by smoothing, and, if needful, stamping a sufficient space of ground, and enclosing it in boards 14 inches wide, set on edge. Into this bed the partially dried peat is thrown, and, as it cracks on the surface by drying, it is compressed by blows with a heavy mallet or flail, or by treading it with flat boards, attached to the feet, somewhat like snow shoes. By this treatment the mass is reduced to a continuous sheet of less than one-half its first thickness, and becomes so firm, that a man's step gives little impression in it. The boards are now removed, and it is cut into blocks by means of a very thin, sharp spade. Every other block being lifted out and placed crosswise upon those remaining, air is admitted to the whole and the drying goes on rapidly. This kind of peat is usually of excellent quality. In North Germany it is called "Baggertorf," i. e. mud-peat.

* Described and figured in Bulletin de la Societe d'Encouragement, August 1857, p. 513; also Dingler's Polytechnisches Journal, Bd. 146, S. 252.

Peat is sometimes dredged by machinery, as will be noticed hereafter.

12.—*The Moulding of Peat.*

When black, earthy or pitchy peat cannot be cut, and is not so saturated with water as to make a mud ; it is, after raking or picking out roots, etc., often worked into a paste by the hands or feet, with addition of water, until it can be formed into blocks which, by slow drying, acquire great firmness. In Ireland this product is termed "hand-peat." In Germany it is called "Formtorf," *i. e.* moulded peat, or "Backtorf," *i. e.* baked peat.

The shaping is sometimes accomplished by plastering the soft mass into wooden moulds, as in making bricks.

13.—*Preparation of Peat Fuel by Machinery, etc.*

Within the last 15 years, numerous inventions have been made with a view to improving the quality of peat fuel, as well as to expedite its production. These inventions are directed to the following points, viz.: 1. *Condensation* of the peat, so as bring more fuel into a given space, thus making it capable of giving out an intenser heat; at the same time increasing its hardness and toughness, and rendering it easier and more economical of transportation. 2. *Drying* by artifical heat or reducing the amount of water from 20 or 25 *per cent.* to half that quantity or less. This exalts the heating power in no inconsiderable degree. 3. *Charring.* Peat-charcoal is as much better than peat, for use where intense heat is required, as wood charcoal is better than wood. 4. *Purifying from useless matters.* Separation of earthy admixtures which are incombustible and hinder draught.

A.—Condensation by Pressure.

Pressing Wet Peat.—The condensation of peat was first attempted by subjecting the fresh, wet material, to severe pressure. As long ago as the year 1821, Pernitzsch, in Saxony, prepared peat by this method, and shortly afterwards Lord Willoughby d'Eresby, in Scotland, and others, adopted the same principle. Simple pressure will, indeed, bring fresh peat at once into much smaller bulk; but, if the peat be fibrous and light, and for this reason require condensation, it is also elastic, and, when the pressure is relieved, it acquires again much of its original volume.

Furthermore, although pressure will squeeze out much water from a saturated well-ripened peat, the complete drying of the pressed blocks usually requires as much or more time than that of the unpressed material, on account of the closeness of texture of the surface produced by the pressure.

The advantages of subjecting fresh peat to pressure in the ordinary presses, it is found, are more than offset by the expense of the operation, and it is therefore unnecessary to give the subject further attention.

Fresh peat appears however to have been advantageously pressed by other mechanical means. Two methods require notice.

Mannhardt's Method, invented about the year 1858, has been practically applied on the large scale at *Schleissheim*, Bavaria. Mannhardt's machine consists of two colossal iron rolls, each of 15 feet diameter, and 6½ feet length, geared into each other so as to revolve horizontally in opposite directions and with equal velocity. These rolls are hollow, their circumference consists of stout iron plate perforated with numerous small holes, and is supported by iron bars which connect the ends of the roll, having intervals between them of about one inch. Each roll is covered by an endless band of hair cloth, stretched over and kept in place by rollers. The rolls are operated by a steam engine of 12 horse power. The fresh peat is thrown into a hopper, and passing between the rolls, loses a considerable share of its water, issuing as a broad continuous sheet, which is divided into blocks by an arrangement presently to be described. The cloth, covering the rolls, must have great strength, sufficient porosity to allow water to pass it freely, and such closeness of texture as to retain the fine particles of peat. Many trials have led to the use of a fabric, specially made for the purpose, of goat's hair. The cloth for each pair of rolls, costs $160.

The peat at Schleissheim is about 5 feet in depth, and consists of a dark-brown mud or paste, free from stones and sticks, and penetrated only by fine fibers. The peat is thrown up on the edge of a ditch, and after draining, is moved on a tram-way to the machine. It is there thrown upon a chain of buckets, which deliver it at the hopper above the rolls. The rolls revolve once in $7\frac{1}{3}$ minutes and at each revolution turn out a sheet of peat, which cuts into 528 blocks. Each block has, when moist, a length of about 12 inches, by 5 inches of width and $1\frac{1}{4}$ inches of thickness, and weighs on the average $1\frac{1}{2}$ lbs. The water that is pressed out of the peat, falls within the rolls and is conducted away; it is but slightly turbid from suspended particles. The band of pressed peat is divided in one direction as it is formed, by narrow slats which are secured horizontally to the press-cloth, at about 5 inches distance from each other. The further division of the peat is accomplished by a series of six circular saws, under which the peat is carried as it is released from the rolls, by a system of endless cords strung over rollers. These cords run parallel until the peat passes the saws; thenceforth they radiate, so that the peat-blocks are separated somewhat from each other. They are carried on until they reach a roll, over which they are delivered upon drying lattices. The latter move regularly under the roll; the peats arrange themselves upon them edgewise, one leaning against the other, so as to admit of free circulation of air. The lattices are loaded upon cars, and moved on a tram-way to the drying ground, where they are set up in frames.

The peat-cake separates well from the press-cloths; but the pores of the latter become somewhat choked by fine particles that penetrate them. They are therefore washed at each revolution by passing before a pipe from which issue, against them, a number of jets of water under high pressure. The blocks, after leaving the machine, are soft, and require 5 or 6 days to become air-dry. When dry they are dense and of good quality, but not better than the same raw material yields by simple moulding. The capacity of the rolls, which easily turn out 100,000 peats in 24 hours, greatly exceeds at present that of the drying arrangements, and for this reason the works are not, as yet, remunerative. The rolls are, in reality, a simple forming machine. The pressure they exert on the peat, is but inconsiderable, owing to its soft pasty character; and since the pair of rolls costs $8000 and can only be worked 3 to 4 months, this method must be regarded rather as an ingenious and instructive essay in the art of making peat-fuel, than as a practical success. The persevering efforts of the inventor may yet overcome all difficulties and prove the complete efficacy of the method. It is especially important, that blocks of greater thickness should be produced, since those now made, pack together too closely in the fire.

Neustadt Method.—At Neustadt, in Hanover, a loose-textured fibrous peat was prepared for metallurgical use in 1860, by passing through iron rolls of ordinary construction. The peat was thereby reduced two-thirds in bulk, burned more regularly, gave a coherent coal, and withstood carriage better. The peat was, however, first cut into sods of regular size, and these were fed into the rollers by boys.

(*b*) *Pressing Air-dried Peat.*

Some kinds of peat, when in the air-dry and pulverized state, yield by great pressure very firm, excellent, and economical fuel.

Lithuanian Process.—In Lithuania, according to Leo,* the following method is extensively adopted. The bog is drained, the surface moss or grass-turf and roots are removed, and then the peat is broken up by a simple spade-plow, in furrows 2 inches wide and 8 or 10 inches deep. The broken peat is repeatedly traversed with wooden harrows, and is thus pulverized and dried. When suitably dry, it is carried to a magazine, where it is rammed into moulds by a simple stamp of two hundred pounds weight. The broken peat is reduced to two-fifths its first bulk, and the blocks thus formed are so hard, as to admit of cutting with a saw or ax without fracture. They require no further drying, are of a deep-brown color, with lustrous surfaces, and their preparation may go on in winter with the stock of broken peat, which is accumulated in the favorable weather of summer. In this manufacture there is no waste of material.

The peat is dry enough for pressing when, after forming in the hands to a ball, it will not firmly retain this shape, but on being let fall to the ground, breaks to powder. The entire cost of preparing 1000 peats for use, or market, was 2 Thalers, or $1.40. Thirty peats, or "stones" as they are called from their hardness, have the bulk of two cubic feet, and weigh 160 lbs. The cost of preparing a hundred weight, was therefore, (in 1859,) four Silvergroschen, or about 10 cents.

The stamp is of simple construction, somewhat like a pile-driver, the mould and face of the ram being made of cast iron. The above process is not applicable to ***fibrous peat.***

* Berg- und Huettenmænnische Zeitung, 1859, Nr. 26.

(*c*) *Pressing Hot-dried Peat.*

The two methods to be next described, are similar to the last mentioned, save that the peat is *hot-pressed.*

Gwynne's Method.—In 1853, Gwynne of London, patented machinery and a method for condensing peat for fuel. His process consisted, first, in rapidly drying and pulverizing the fresh peat by a centrifugal machine, or by passing between rollers, and subsequent exposure to heat in revolving cylinders; and, second, in compressing the dry peat-powder in a powerful press at a high temperature, about 180° F. By this heat it is claimed, that the peat is not only thoroughly dried, but is likewise partially decomposed; *bituminous matters being developed, which cement the particles to a hard dense mass.* Gwynne's machinery was expensive and complicated, and although an excellent fuel was produced, the process appears not to have been carried out on the large scale with pecuniary success.

A specimen of so-called "Peat coal" in the author's possession, made in Massachusetts some years ago, under Gwynne's patent, appears to consist of pulverized peat, prepared as above described; but contains an admixture of rosin. It must have been an excellent fuel, but could not at that time compete with coal in this country.

*Exter's Method.**

In 1856, Exter, of Bavaria, carried into operation on an extensive scale, a plan of preparing peat-fuel in some respects not unlike the last mentioned method. Exter's works, belonging to the Bavarian Government, are on the Haspelmoor, situated between Augsburg and Munich. According to Ruehlmann, who examined them at the

* Henneberg's Journal fuer Landwirthschaft, 1858, S. 42.

command of the Hanoverian Government in 1857, the method is as follows:—1. The bog is laid dry by drains and the surface is cleared of bushes, roots, and grass-turf, down to good peat. 2. The peat is broken up superficially to the depth of about one inch, by a gang of three plows, propelled by a portable steam engine. 3. The peat is further pulverized by a harrow, drawn by a yoke of oxen. 4. In two or three days after harrowing, the peat is turned by an implement like our cultivator, this process being repeated at suitable intervals. 5. The fine and air-dry peat is gathered together by scrapers, and loaded into wagons; then drawn by rope connected with the engine, to the press or magazine. 6. If needful, the peat, thus collected, is further pulverized by passing it through toothed rollers. 7. The fine peat is now introduced into a complicated drying oven, see figures 5 and 6. It falls through the opening *T*, and is moved by

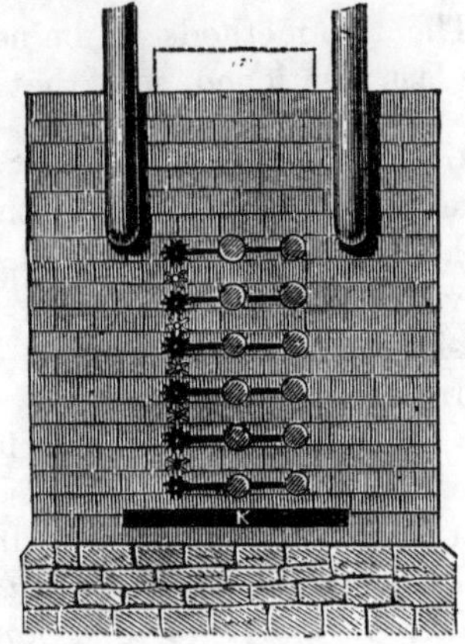

Fig. 5.—EXTER'S DRYING OVEN.

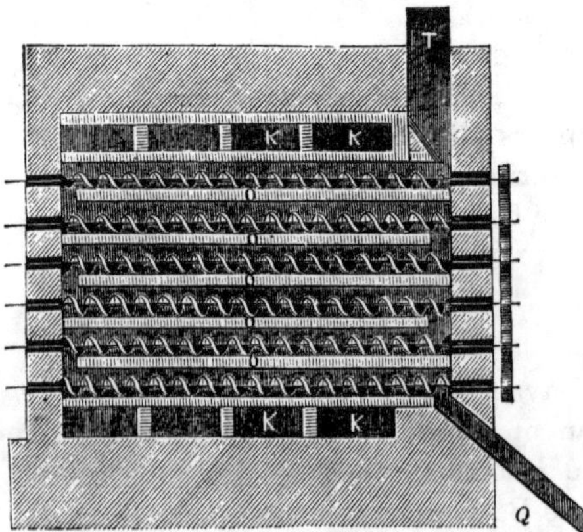

Fig. 6.—EXTER'S DRYING OVEN.

means of the spirals along the horizontal floors *O*, *O*, falling from one to another until it emerges at *Q*. The floors, *O*, *O*, are made by wide and thin iron chambers, through which passes waste steam from an engine. The oven is heated further by hot air, which circulates through the canals *K*,*K*. The peat occupies about one hour in its passage through the oven and falls from *Q*, into the press, having a temperature of from 120° to 140° Fahrenheit. The press employed at Staltach is essentially the same as that now used at the Kolbermoor, and figured on p. 125. It is a powerful eccentric of simple construction, and turns out continuously 40 finished peats per minute. These occupy about one-fourth the space of the peat before pressing, the cubic foot weighing about 72 lbs. The peats are 7 inches long, 3 inches wide, and one half to three quarters of an inch thick, each weighing three quarters of a pound. Three presses furnish annually 180,000 cwt. of condensed peat, which is used exclusively for firing locomotives. Its specific gravity is 1.14, and its quality as fuel is excellent. Ruehlmann estimated its cost, at Haspelmoor in 1857, at 8½ Kreuzers, or a little more than 6 cents per cwt., and calculated that by adopting certain obvious improvements, and substituting steam power for the labor of men and cattle, the cost might be reduced to 6½ Kreuzers, or a little more than 4 cents per cwt.

Exter's method has been adopted with some modifications at Kolbermoor, near Munich, in Bavaria, at Miskolz, in Hungary, and also at the Neustadt Smelting Works, in Hanover. At the latter place, however, it appears to have been abandoned for the reasons that it could be applied only to the better kinds of peat; and the expense was there so great, that the finished article could not compete with other fuel in the Hanoverian markets.

Details of the mechanical arrangements at present employed on the Kolbermoor, are as follows: After the bog is drained, and the surface cleared of dwarf pines, etc., and suitably leveled, the peat is plowed by steam. This is accomplished in a way which the annexed cut serves to illustrate. The plot to be plowed, is traversed through the middle by the rail-way *x,y*. A locomotive *a*, sets in motion an endless wire-rope, which moves upon large

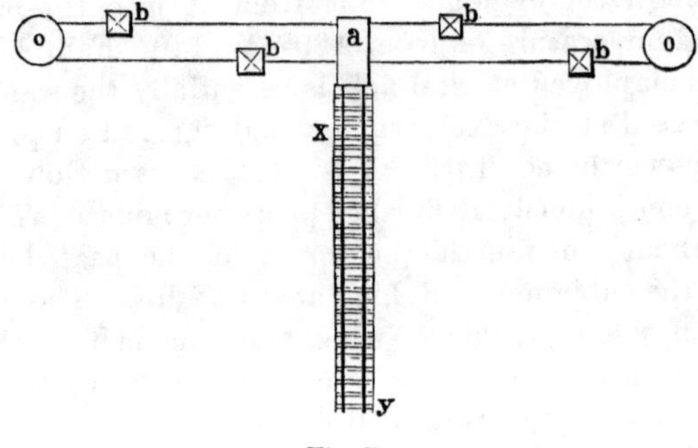

Fig. 7.

horizontal pulleys *o,o*, stationed at either border of the land. Four gang plows *b,b*, are attached to the rope, and as the latter is set in motion, they break up the strip of peat they pass over, completely. The locomotive and the pulleys are then moved back, and the process is repeated until the whole field has been plowed. The plows are square frames, carrying six to eight shares and as many coulters.

The press employed at Kolbermoor, is shown in figs. 8 and 9. The hot peat falls into the hopper, *b,c*. The plunger *d*, worked in the cavity *e*, by an eccentric, allows the latter to fill with peat as it is withdrawn, and by its advance compresses it into a block. The blocks *m*, once formed, by their friction in the channel *e*, oppose enough resistance to the peat to effect its compression. In order to regulate this resistance according to the varying quality of the peat, the piece of metal *g*, which hangs on a pivot at *o*, is depressed or raised, by the screw *i*, so as to contract or

Fig. 8.—EXTER'S PEAT PRESS.

enlarge the channel. At each stroke of the plunger a block is formed, and when the channel *e* is once filled, the peats fall continuously from its extremity. Their dimensions are 7 inches long, $3\frac{1}{2}$ wide, and $1\frac{1}{2}$ thick.

Several presses are worked by the same engine at the Kolbermoor, each of which turns out daily 200 to 300 cwt. of peats, which, in 1863, were sold at 24 kreuzers (16 cents), per cwt.

Fig. 9.—EXTER'S PEAT PRESS.

C. Hodgson has patented in Great Britain a compressing-ram similar to Exter's, and works were put up at Derrylea, in Ireland, some years ago, in which Exter's process of manufacturing peat fuel appears to have been adopted.

Elsberg's Process.

Dr. Louis Elsberg, of New York City, has invented a modification of Exter's method, which appears to be of great importance. His experimental machine, which is in operation near Belleville, N. J., consists of a cylindrical pug-mill, in which the peat, air-dried as in Exter's method, is further broken, and at the same time is subjected to a current of steam admitted through a pipe and jacket surrounding the cylinder. The steamed peat is then condensed by a pair of presses similar to that just described, which are fed directly from the mill. In this way the complicated drying oven of Exter is dispensed with. Elsberg & Co. are still engaged in perfecting their arrangements. Some samples of their making are of very excellent quality, having a density of 1.2 to 1.3.

The pressing of air-dry peat only succeeds when it is made warm, and is, at the same time, moist. In Exter's original process the peat is considerably dried in the ovens, but on leaving them, is so moist as to bedew the hand that is immersed in it. It is, in fact, steamed by the vaporization of its own water. In Elsberg's process, the air-dry peat is not further desiccated, but is made moist and warm by the admission of hot steam. The latter method is the more ready and doubtless the more economical of the two. Whether the former gives a dryer product or not, the author cannot decide. Elsberg's peat occurs in cylindrical cakes 2 inches broad, and one inch in thickness. The cakes are somewhat cracked upon the edges, as if by contraction, in drying. When wet, the surface of the cakes swells up, and exfoliates as far as the water has penetrated. In the fire, a similar breaking away of the surface takes place, and when coked, the coal is but moderately coherent.

The reasons why steamed peat admits of solidification by pressure, are simply that the air, ordinarily adhering to the fibres and particles, is removed, and the fibres themselves become softened and more plastic, so that pressure brings them into intimate contact. The idea that the heat developes bituminous matters, or fuses the resins which exist in peat, and that these cement the particles, does not harmonize with the fact that the peat, thus condensed, flakes to pieces by a short immersion in water.

The great advantage of Exter's and Elsberg's method consists in avoiding what most of the others require, viz.: the expensive transportation and handling of fresh peat, which contains 80 to 90 *per cent.* of water, and the rapid removal of this excess of water before the manufacture. In the other methods the surplus water must be slowly removed during or after condensation.

Again, enough peat may be air-dried and stored during summer weather, to supply a machine with work during the whole year.

Its disadvantages are, that it requires a large outlay of capital and great expenditure of mechanical force. Its product is, moreover, not adapted for coking.

B.—Condensation without Pressure.

The methods of condensing peat, that remain to be described, are based upon radically different principles from those already noticed. In these, little or no pressure is employed in the operations; but advantage is taken of the important fact that when wet or moist peat is ground, cut or in any way reduced to a pulpy or pasty consistence, with destruction of the elastic fibres, it will, on drying, shrink together to a coherent mass, that may acquire a density and toughness much greater than it is possible to obtain by any amount of mere pressure.

The various processes that remain to notice are essentially reducible to two types, of which the French method, invented by Challeton, and the German, invented it appears by Weber, are the original representatives. The former method is only applicable to earthy, well-decomposed peat, containing little fibre. The latter was originally applied to fibrous moss-peat, but has since been adapted to all kinds. Other inventors, English, German, and American, have modified these methods in their details, or in the construction of the requisite machinery, rendering them more perfect in their execution and perhaps more profitable in their results; but, as regards the essential principles of production, or the quality of product, no advance appears to have been made beyond the original inventors.

(a) *Condensation of Earthy Peat.*

Challeton's Method consists essentially in destroying the fibres, and reducing the peat by cutting and grinding with water to a pulp; then slowly removing the liquid, until the peat dries away to a hard coherent mass. It provides also for the purification of the peat from earthy matters. It is, in many respects, an imitation of the old Dutch and Irish mode of making "hand peat" (*Baggertorf*), and is very like the paper manufacture in its operations. Challeton's Works, situated near Paris, at Mennecy, near Montanges, were visited in 1856 by a Commission of the Agricultural Society of Holstein, consisting of Drs. Meyn and Luetkens, and also by Dr. Ruehlmann, in the interest of the Hanoverian Government. From their account* the following statements are derived.

The peat at Mennecy comes from the decay of grasses, is black, well decomposed, and occasionally intermingled with shells and sand. The moor is traversed by canals, which serve for the transport of the excavated peat in boats. The peat, when brought to the manufactory, is emptied into a cistern, which, by communicating with the adjacent canal, maintains a constant level of water. From

* Henneberg's Journal fuer Landwirthschaft, 1858, p.p. 42 and 83.

this cistern the peat is carried up by a chain of buckets and emptied into a hopper, where it is caught by toothed cylinders in rapid revolution, and cut or torn to pieces. Thence it passes into a chamber where the fine parts are separated from unbroken roots and fibres by revolving brushes, which force the former through small holes in the walls of the chamber, while the latter are swept out through a larger passage. The pulverized peat finally falls into a cistern, in which it is agitated by revolving arms. A stream of water constantly enters this vessel from beneath, while a chain of buckets as rapidly carries off the peat pulp. All sand, shells, and other heavy matters, remain at the bottom of this cistern.

The peat pulp, thus purified, flows through wooden troughs into a series of basins, in which the peat is formed and dried. These basins are made upon the ground by putting up a square frame (of boards on edge,) about one foot deep, and placing at the bottom old matting or a layer of flags or reeds. Each basin is about a rod square, and 800 of them are employed. They are filled with the peat pulp to the top. In a few days the water either filters away into the ground, or evaporates, so that a soft stratum of peat, about 3 inches in thickness, remains. Before it begins to crack from drying, it is divided into blocks, by pressing into it a light trellis-like framework, having thin partitions that serve to indent the peat in lines corresponding to the intended divisions. On further drying, the mass separates into blocks at the lines thus impressed, and in a few days, they are ready to remove and arrange for further desiccation.

The finished peats from Challeton's works, as well as those made by the same method near Neuchatel, Switzerland, by the Messrs. Roy, were of excellent quality, and in the opinion of the Commission from Holstein, the method is admirably adapted for the purification and concentration of the heavy kinds of peat.

In Holstein, a French company constructed, and in 1857 worked successfully a portable machine for preparing peat on this plan, but were shortly restrained by legal proceedings. Of their later operations we have no information.

No data are at hand regarding the cost of producing fuel by Challeton's machinery. It is believed, however, that his own works were unremunerative, and several manufactories on his pattern, erected in Germany, have likewise proved unprofitable. The principle is, however, a good one, though his machinery is only applicable to earthy or pitchy, and not to very fibrous peat. It has been elsewhere applied with satisfactory results.

Simplified machinery for applying Challeton's method is in operation at Langenberg, near Stettin, in Prussia.* The moss-meadows along the river Oder, near which Langenberg is situated, are but a foot or so higher at the surface than the medium level of this river, and are subject to frequent and sudden inundations, so that draining and partial drying of the peat are out of the question. The character of the peat is unadapted to cutting by hand, since portions of it are pitchy and crumble too easily to form good sods ; and others, usually the lower layers, at a depth of seven feet or more, are made up to a considerable extent of quite firm reeds and flags, having the consistence of half decayed straw. The earthy peat is manufactured after Challeton's method. It is raised with a steam dredger of 20 horse power, and emptied into flat boats, seven in number, which are drawn to the works by an endless rope operated by horse power. The works themselves are situated on a small sand hill in the middle of the moor, and communicate by canal with the dredger and with the drying

* Dingler's Journal, Oct., 1864.

ground. A chain of buckets, working in a frame 45 feet long, attached by a horizontal hinge to the top of the machine house, reaches over the dock where the boats haul up, into the rear end of the latter; and, as the buckets begin to raise the peat, the boat itself is moved under the frame towards the house, until, with a man's assistance, its entire load is taken up. The contents of one boat are six square yards, with a depth of one foot, and a boat is emptied in 20 minutes time. Forty to forty-four boat-loads are thus passed into the pulverizing machine daily, by two chains of buckets.

The peat-mud falls from the buckets into a large wooden trough, which branches into two channels, conducting to two large tubs standing side by side. These tubs are 10 feet in diameter and 2 feet deep, and are made of 2-inch plank. Within each tub is placed concentrically a cylindrical sieve, or colander, 8 feet in diameter and 2 feet high, made of $\frac{3}{8}$ round iron, and it is within this that the peat is emptied. The peat is stirred and forced through the meshes of the sieve by four arms of a shaft that revolves 20 times per minute, the arms carrying at their extremities stiff vertical brooms, which rub the inside of the sieve.

In these four tubs the peat is pulverized under addition of water; the fine parts pass the sieves, while the latter retain the coarse fibres, roots, etc. The peat-mud flows from the tubs into mills, made like a flour mill, but the "stones" constructed of hard wood. The "stones" have a diameter of 8 feet 6 inches; the lower is 8 inches; the upper 21 inches thick. The pressure of the upper "stone" is regulated by adjusting the level of the discharging channel, so that the "stone" may be more or less buoyed, or even fully floated by the water with which it is surrounded.

The peat-substance, which is thus finely ground, gathers from the four mills into a common reservoir whence it is lifted by a centrifugal pump into a trough, which distributes it over the drying ground.

The drying ground consists of the surface formed by grading the sand hill, on which the works are built, and includes about 30 English acres. This is divided into small plots, each of which is enclosed on three sides with a wall of earth, and on the fourth side by boards set on edge. Each plot is surrounded by a ditch to carry off water, and by means of portable troughs, the peat is let on from the main channel. The peat-slime is run into these beds to the depth of 20 to 22 inches, an acre being covered daily. After 4 to 8 days, according to the weather, the peat has lost so much water, which rapidly soaks off through the sand, that its surface begins to crack. It is then thoroughly trodden by men, shod with boards 5 inches by 10 inches, and after 6 to 8 days more, it is cut with sharp spades into sods. The peats are dried in the usual manner.

The works at Langenberg yielded, in 1863, as the result of the operations of 60 days of 12 hours each, 125,000 cwt. of marketable peat. It is chiefly employed for metallurgical purposes, and sells at 3⅓ Silvergroschen, or nearly 8 cents per cwt. The specific gravity of the peat ranges from 0.73 to 0.90.

Roberts' Process.

In this country attempts have been made to apply Challeton's method. In 1865, Mr. S. Roberts, of Pekin, N. Y., erected machinery at that place, which was described in the "Buffalo Express," of Nov. 17, 1865, as follows: —

"In outward form, the machine was like a small frame house on wheels, supposing the smoke-stack to be a chimney. The engine and boiler are of locomotive style; the engine being of thirteen horse power. The principal features of the machine are a revolving elevator and a conveyer. The elevator is seventy-five feet long, and runs from the top of the machine to the ground, where the peat is dug up, placed on the elevator, carried to the top of the machine, and dropped into a revolving wheel that cuts it up; separates from it all the coarse particles, bits of sticks, stones, etc.; and throws them to one side. The peat is next dropped into a box below, where water is passed in, sufficient to bring it to the consistency of mortar. By means of a slide under the control of the engineer, it is next sent to the rear of the machine, where the conveyer, one hundred feet long, takes it, and carries it within two rods of the end; at which point the peat begins to drop through to the ground to the depth of about four or five inches. When sufficient has passed through to cover the ground to the end of the conveyer,—two rods,—the conveyer is swung around about two feet, and the same process gone through, as fast as the ground under the elevator, for the distance of two rods in length and two feet in width gets covered, the elevator being moved. At each swing of the elevator, the peat just spread is cut into blocks (soft ones, however) by knives attached to the elevator. It generally takes from three to four weeks before it is ready for use. It has to lie a week before it is touched, after the knives pass through it; when it is turned over, and allowed to lie another week. It has then to be taken up, and put in a shed, and within a week or ten days can be used, although it is better to let it remain a little longer time. The machine can spread the peat over eighteen square rods of ground — taking out one square rod of peat — without being moved. After the eighteen rods are covered, the machine is moved two rods ahead, enabling it to again spread a semicircular space of some thirty-two feet in width by eighteen rods in length. The same power, which drives the engine, moves the machine. It is estimated by Mr. Roberts, that, by the use of this machine, from twenty to thirty tons of peat can be turned out in a day."

Mr. Roberts informs us that he is making (April 1866,) some modifications of his machinery. He employs a revolving digger to take up the peat from the bed, and carry it to the machine. At the time of going to press, we do not learn whether he regards his experiments as leading to a satisfactory conclusion, or otherwise.

Siemens' method.

Siemens, Professor of Technology, in the Agricultural Academy, at Hohenheim, successfully applied the following mode of preparing peat for the Beet Sugar Manufactory at Bœblingen, near Hohenheim, in the year 1857. Much of the peat there is simply cut and dried in the usual manner. There is great waste, however, in this process, owing to the frequent occurrence of shells and clay, which destroy the coherence of the peat. Besides, a large quantity of material accumulates in the colder months, from the ditches which are then dug, that cannot be worked in the usual manner at that time of the year. It was to economize this otherwise useless material that the following process was devised, after a failure to employ Challeton's method with profit.

In the first place, the peat was dumped into a boarded cistern, where it was soaked and worked with water, until it could be raised by a chain of buckets into the pulverizer.

The pulverization of the peat was next effected by passing it through a machine invented by Siemens, for pulping potatoes and beets. This machine, (the same we suppose as that described and figured in Otto's Landwirthschaftliche Gewerbe), perfectly breaks up and grates the peat to a fine pulp, delivers it in the consistency of mortar into the moulds, made of wooden frames, with divisions to form the peats. The peat-paste is plastered by hand into these moulds, which are immediately emptied to fill again, while the blocks are carried away to the drying ground where they are cured in the ordinary style without cover.

In this simple manner 8 men were able to make 10,000 peats daily, which, on drying, were considerably denser and harder than the cut peat.

The peat thus prepared, cost about one-third more than the cut peat. Siemens reckoned, this greater cost would be covered by its better heating effect, and its ability to withstand transportation without waste by crumbling.

(*b*) *Condensation of fibrous peat.*

Weber's method.

At Staltach, in Southern Bavaria, Weber has established an extensive peat works, of which Vogel has given a circumstantial account.* The peat at Staltach is very light and fibrous, but remarkably free from mineral matters, containing less than 2 *per cent.* of ash in the perfectly dry substance. The moor is large, (475 acres), and the peat is from 12 to 20 feet in depth. The preparation consists in converting the fresh peat into pulp or paste, forming it into moulds and drying it; at first by exposure to the air at ordinary temperature, and finally, by artificial heat, in a drying house constructed for the purpose.

The peat is cut out by a gang of men, in large masses, cleared of coarse roots and sticks, and pushed on tram

* Dingler's Polytechnisches Journal, Bd. 152, S. 272. See also, Knapp, Lehrbuch der Chemischen Technologie, 3te Auflage, 1., 167.

wagons to the works, which are situated lower than the surface of the bog. Arrived at the works, the peat is carried upon an inclined endless apron, up to a platform 10 feet high, where a workman pushes it into the pulverizing mill, the construction of which is seen from the accompanying cut. The vertical shaft *b* is armed with

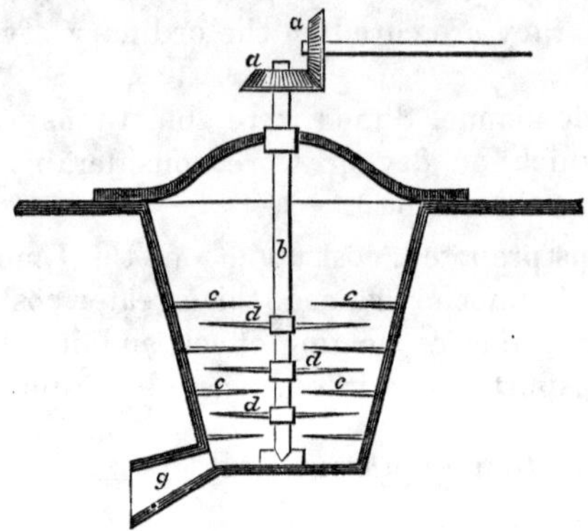

Fig. 10.—WEBER'S PEAT MILL.

sickle-shaped knives, *d*, which revolve between and cut contrary to similar knives *c*, fixed to the interior of the vessel. The latter is made of iron, is 3½ feet high, 2 feet across at top and 1½ feet wide at the bottom. From the base of the machine at *g*, the perfectly pulverized or minced peat issues as a stiff paste. If the peat is dry, a little water is added. Vogel found the fresh peat to contain 90 *per cent.* of water, the pulp 92 *per cent.* Weber's machine, operated by an engine of 10 horse power, working usually to half its capacity only, reduced 400 cubic feet of peat per hour, to the proper consistency for moulding.

Three modes of forming the paste into blocks have been practiced. One was in imitation of that employed with mud-peat. The paste was carried by railway to sheds, where it was filled by hand into moulds 17 inches by $7\frac{1}{4}$ by $5\frac{1}{2}$ inches, and put upon frames to dry. These sheds occupied together 52,000 square feet, and contained at once 200,000 peats. The peats remained here 8 to 14 days or more, according to the weather, when they were either removed to the drying house, or piled in large stacks to dry slowly out-of-doors. The sheds could be filled and emptied at least 12 times each season, and since they protected from light frosts, the season began in April and lasted until November.

The second mode of forming the peat was to run off the pulp into large and deep pits, excavated in the ground, and provided with drains for carrying off water. The water soaked away into the soil, and in a few weeks of good weather, the peat was stiff enough to cut out into blocks by the spade, having lost 20 to 25 *per cent.* of its water, and 15 *per cent.* of its bulk. The blocks were removed to the drying sheds, and set upon edge in the spaces left by the shrinking of the peats made by the other method. The working of the peat for the pits could go on, except in the coldest weather, as a slight covering usually sufficed to protect them from frost.

Both of these methods have been given up as too expensive, and are replaced, at present, by the following:

In the third method the peat-mass falls from the mill into a hopper, which directs it between the rolls *A B* of fig. 11, (see next page). The roll *A* has a series of boxes on its periphery *m m*, with movable bottoms which serve as moulds. The peat is carried into these boxes by the rolls *c c*. The iron projections *n n* of the large roll *B*, which work cog-like into the boxes, compress the peat gently and, at last, the eccentric *p* acting upon the pin *z*, forces up the movable bottom of the box and throws out the peat-block upon an endless band of cloth, which carries it to the drying place.

The peats which are dried at first under cover and therefore slowly, shrink more evenly and to a greater extent than those which are allowed to dry rapidly. The latter become cracked upon the surface and have cavities internally, which the former do not. This fact is of great importance for the density of the peat, for its usefulness in producing intense heat, and its power to withstand carriage.

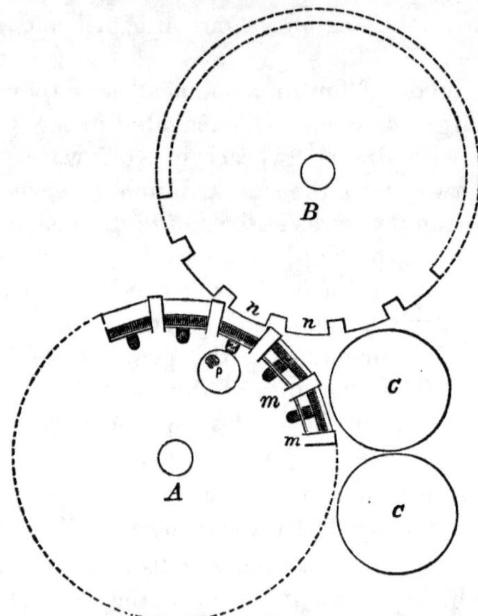

Fig. 11—WEBER'S PEAT MOULDING MACHINE.

The *complete drying* is, on the other hand, by this method, a much slower process, since the dense, fissureless exterior of the peats hinders the escape of water from within. It requires, in fact, several months of ordinary drying for the removal of the greater share of the water, and at the expiration of this time they are still often moist in the interior.

Artificial drying is therefore employed to produce the most compact, driest, and best fuel.

Weber's *Drying-house* is 120 feet long and 46 feet wide. Four large flues traverse the whole length of it, and are heated with the pine roots and stumps which abound in the moor. These flues are enclosed in brick-work, leaving a narrow space for the passage of air from without, which is heated by the flues, and is discharged at various openings in the brick-work into the house itself, where the peat is arranged on frames. The warm air being light, ascends through the peat, charges itself with moisture, thereby becomes heavier and falls to the floor, whence it is drawn off by flues of sheet zinc that pass up through the roof. This house holds at once 300,000 peats, which are heated to 130° to 145° F., and require 10 to 14 days for drying.

The effect of the hot air upon the peat is, in the first place, to soften and cause it to swell; it, however, shortly begins to shrink again and dries away to masses of great solidity. It becomes almost horny in its character, can be broken only by a heavy blow, and endures the roughest handling without detriment. Its quality as fuel is correspondingly excellent.

The effects of the mechanical treatment and drying on the Staltach peat, are seen from the subjoined figures:

	Specific Gravity.	*Lbs. per Cubic Foot.*	*Per cent. of Water.*
Peat, raised and dried in usual way,	0.24	15	18 to 20
Machine-worked and hot-dried	0.65	35	12

Vogel estimates the cost of peat made by Weber's method at 5 Kreuzers per (Bavarian) hundred weight, while that of ordinary peat is 13½ Kreuzers. Schrœder, in his comparison of machine-wrought and ordinary peat, demonstrates that the latter can be produced much cheaper than was customary in Bavaria, in 1859, by a better system of labor.

Weber's method was adopted with some improvements in an extensive works built in 1860, by the Government of Baden, at Willaringen, for the purpose of raising as much fuel as possible, during the course of a lease that expired with the year 1865.

*Gysser's method.**—Rudolph Gysser, of Freiburg, who

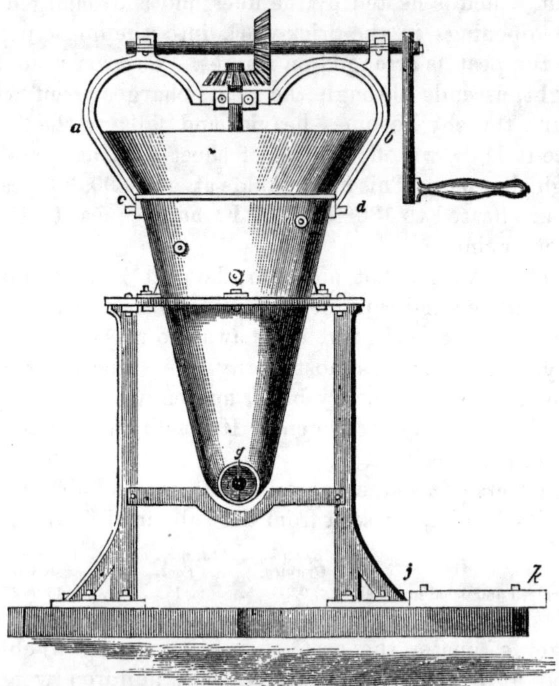

Fig. 12.—GEYSSER'S PEAT MACHINE.

was charged with the erection of the works at Willaringen just alluded to, invented a portable hand-machine on

* Der Torf; seine Bildung und Bereitungsweise, von Rudolph Gysser, Weimar, 1864.

the general plan of Weber, but with important improvements; and likewise omitted and varied some details of the manufacture, bringing it within the reach of parties of small means.

In the accompanying cuts, (figs. 12, 13, and 14), are given an elevation of Gysser's machine, together with a bird's-eye view and vertical section of the interior mechanism.

It consists of a cast iron funnel *c d i* of the elevation, (fig. 12), having above a sheet iron hopper *a b* to receive the peat, and within a series of six knives fastened in a spiral,

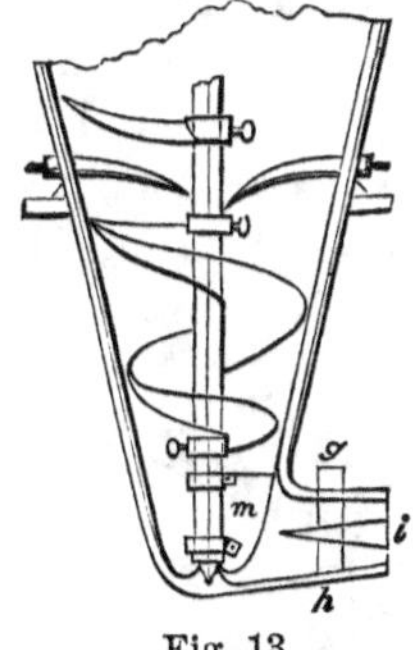

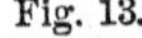
Fig. 13.

Fig. 14.

and curving outwards and downwards, (figs. 13 and 14); another series of three similar knives is affixed to a vertical shaft, which is geared to a crank and turned by a man standing on the platform *j k;* these revolving knives curve upwards and cut between and in a direction contrary to the fixed knives; below the knives, and affixed to the shaft a spiral plate of iron and a scraper *m*, (fig. 13), serve to force the peat, which has been at once minced and carried downwards by the knives, as a somewhat compressed mass through the lateral opening at the bottom of the funnel, whence it issues as a continuous hollow cylinder like drain-tile, having a diameter of four inches. The iron cone *i*, held in the axis of the opening by the thin and sharp-edged support *g h*, forms the bore of the tube of peat as it issues. Two men operate the machine; one turning the crank, which, by suitable gearing, works the shaft, and the other digging and throwing in the peat. The mass, as it issues from the machine, is received by two boys alternately, who hold below the opening a semi-cylindrical tin-plate shovel, (fig. 15), of the width and length of the required peats, and break or rather wipe them off, when they reach the length of 14 inches.

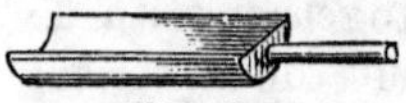

Fig. 15.

The formed peats are dried in light, cheap and portable houses, Fig. 17, each of which consists of six rectangular frames supported one above another, and covered by a light roof. The frames, Fig. 16, have square posts at each corner like a bedstead, and are made by nailing light strips to these posts. The tops of these posts are obtusely beveled to an edge, and at the bottom they are notched to correspond. The direction of the edges and of the notches in two diagonally opposite posts, is at right angles to that of the other two. By this construction the frames, being of the same size, when placed above each other, fit together by the edges and notches of their posts into a structure that cannot be readily overturned. The upper frame has a light shingled roof, which completes the house. Each frame has transverse slats, cast in plaster of Paris, 20 in number, which support the peats. The latter being tubular, dry more readily, uniformly, and to a denser consistence than they could otherwise.

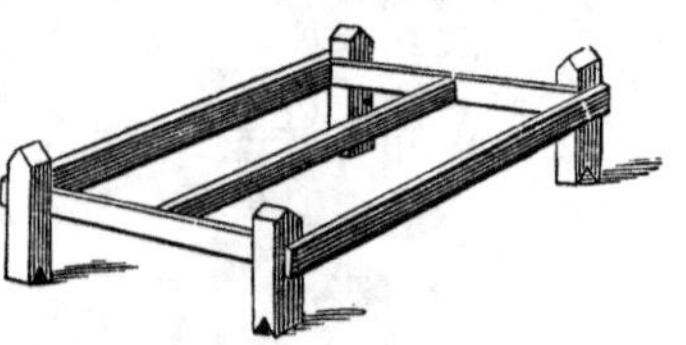

Fig. 16.

The machine being readily set up where the peat is excavated, the labor of transporting the fresh and water-soaked material is greatly reduced. The drying-frames are built up into houses as fast as they are filled from the machine. They can be set up anywhere without difficulty, require no leveling of the ground, and, once filled, no labor in turning or stacking the peats is necessary; while the latter are insured against damage from rain. These advantages, Gysser claims, more than cover their cost.

Fig 17.

The daily production of a machine operated by two men with the assistance of one or two boys, is 2500 to 3000 peats, which, on drying, have 9½ to 10 inches of length, and 2½ in diameter, and weigh, on the average, one pound each.

(c)—*Condensation of peat of all kinds.*—*Weber's method with modified machinery.*

*Schlickeysen's Machine.**—This machine has been in use in Germany since 1860, in the preparation of peat. It appears to have been originally constructed for the working and moulding of clay for making bricks. The principle of its operation is identical with that of Weber's process. The peat is finely pulverized, worked into a homogenous mass, and moulded into suitable forms. Like

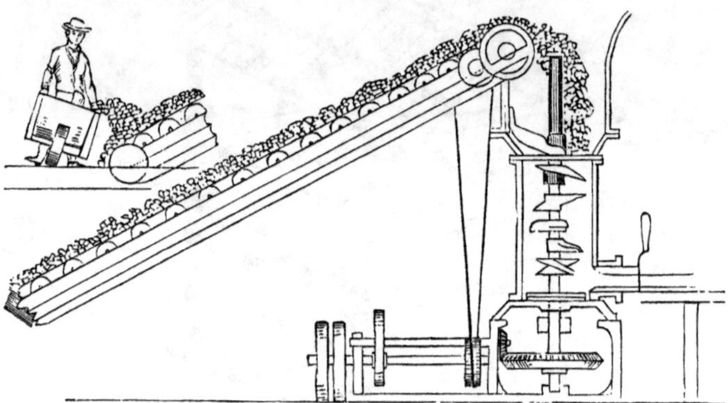

Fig. 18.—SCHLICKEYSEN'S PEAT MILL.

Gysser's machine, it forces the peat under some pressure through a nozzle, or, in the larger kinds through several nozzles, whence it issues in a continuous block or pipe that is cut off in proper lengths, either by hand or by mechanism It consists of a vertical cylinder, through the axis of which revolves a shaft, whereon are fastened the blades, whose edges cut and whose winding figure forces down the peat. The blades are arranged nearly, but not exactly, in a true spiral; the effect is therefore that they act unequally up-

* Dingler's Journal, Bd. 165, S. 184.; und Bd. 172, S. 333.

on the mass, and thus mix and divide it more perfectly. No blades or projections are affixed to the interior of the cylinder. Above, where the peat enters into a flaring hopper, is a scraper, that prevents adhesion to the sides and gives downward propulsion to the peat. The blades are, by this construction, very strong, and not liable to injury from small stones or roots, and effectually reduce the toughest and most compact peat.

Furthermore, addition of water is not only unnecessary in any case, but the peat may be advantageously air-dried to a considerable extent before it enters the machine. Wet peat is, indeed, worked with less expenditure of power; but the moulded peats are then so soft as to require much care in the handling, and must be spread out in single courses, as they will not bear to be placed one upon another. Peat, that is somewhat dry, though requiring more power to work, leaves the machine in blocks that can be piled up on edge and upon each other, six or eight high, without difficulty, and require, of course, less time for curing.

The cut, (fig. 18), represents one of Schlickeysen's portable peat-mills, with elevator for feeding, from which an idea of the pulverizing arrangements may be gathered.

In Livonia, near Pernau, according to Leo, two of Schlickeysen's machines, No. 6, were put in operation upon a purely fibrous peat. They were driven by an engine of 12 horse-power. The peat was plowed, once harrowed, then carted directly to the hopper of the machine. These two machines, with 26 men and 4 horses, produced daily 60,000 peats = 7500 cubic feet. 100 cubic feet of these peats were equal in heating effect to 130 cubic feet of fir-wood, and cost but two-thirds as much. The peats were extremely hard, and dried in a few days sufficiently for use. In 1864, five large Schlickeysen machines were in operation at one establishment at St. Miskolz, in Hungary.

The smaller sizes of Schlickeysen's machine are easily portable, and adapted for horse or hand-power.

*Leavitt's Peat-condensing and Moulding Mill.**—In this country, Mr. T. H. Leavitt, of Boston, has patented machinery, which is in operation at East Lexington, Mass., at the works of the Boston Peat Company. The process is essentially identical with that of Weber, the hot-drying omitted. The fresh peat is pulverized or cut fine, moulded into blocks, and dried on light frames in the open air. The results claimed by Mr. Leavitt, indicate, that his machine is very efficacious.

It consists, principally, of a strong box or cistern, three feet in diameter, and six feet high, the exterior of which, with its gearing, is shown in figure 19. The mill is adapted to be driven by a four horse-power engine.

"The upper portion of the box is divided by a series of horizontal partitions, the upper ones being open lattice-work, and the lower ones perforated with numerous holes. The upright shaft, which rotates in the centre of the box, carries a series of arms or blades, extending alternately on opposite sides, and as these revolve, they cut the peat, and force it through the openings in the diaphragms. The lower portion of the box, in place of complete partitions, has a series of corrugated shelves extending alternately from opposite sides, and the peat is pressed and scraped from these by a series of arms adapted to the work. By this series of severe operations the air-bubbles are expelled from the peat, and it is reduced to a homogeneous paste. When it arrives at the bottom of the box, it is still further compressed by the converging sides of the hopper, and it is received in light moulds which are carried on an endless belt." Mr. Leavitt has patented the

* Scientific American. Feb. 10, 1866; also, Facts about Peat as Fuel. by T. H. Leavitt, 2d Ed., Boston, p. 33.

use of powdered peat for the purpose of preventing the prepared peat from adhering to the moulds.

This mill, it is asserted, will condense 40 tons of crude peat daily, which, at Lexington, is estimated to yield 10 to 14 tons of dry merchantable fuel. The cost of pro-

Fig 19.—LEAVITT'S PEAT MILL.

ducing the latter is asserted to be less than $2.00 per ton; while its present value, in Boston, is $10 per ton. It requires seven men, three boys, and two horses to dig, cart, mill, and spread the peat. The machine costs $600, the needful buildings, engine, etc., from $2000 to $3000. The samples of peat, manufactured by this machine, are of excellent quality. The drying in the open air is said to proceed with great rapidity, eight or ten days being ordinarily sufficient in the summer season. The dry peat, at Lexington, occupies one-fourth the bulk, and has one-fourth to one-third the weight of the raw material; the latter, as we gather, being by no means saturated with water, but well drained, and considerably dry, before milling.

Ashcroft & Betteley's Machinery.

The American Peat Company, of Boston, are the owners of five patents, taken out by Messrs. Ashcroft & Betteley, for peat machinery. They claim to "make fuel equal to the best English Cannel coal," and really do make a very good peat, though with a rather complicated apparatus. The following statement is derived from the circular issued by the company. The machinery consists of the following parts:—

First.—TRITURATING MACHINE—36 inches diameter, 4 feet 6 inches high, with arms both on the inside of this cylinder and on the upright revolving shaft. In the bottom of the cylinder or tub a large slide gate is fitted to work with a lever, so that the peat may be discharged, at pleasure, into the Combing Machine, which is placed directly under this Triturator.

Second.—COMBING MACHINE—Semi-circular vessel 6 feet long and 3 feet 6 inches in diameter. Inside, a shaft is placed, which is provided with fingers, placed one inch apart; the fingers to be 20 inches long, so as to reach within 2 inches of the bottom and sides of this vessel. Another shaft, of the same size and dimensions, is placed at an angle of 45°, 26 inches from the first shaft, with arms of the same dimensions placed upon this shaft, with the same spaces, and so placed that this set of arms pass between the first set, both shafts revolving in the same direction; the second shaft mentioned being driven at double the speed of the first. At the bottom of this Combing Machine is to be fixed a gate, to be operated by a lever, to deliver, at pleasure, the cleansed peat into the Manipulator or Kneeding Machine.

Third.—MANIPULATOR.—A Tube of iron 7 feet long and 16 inches diameter, fitted with a shaft, with flanges upon it, to gain 6 inches in each revolution.

Fourth.—CONVEYOR.—This Conveyor, to be made with two endless chains and buckets of iron, with a driving shaft. The hopper, to receive the peat when first taken from the bog, to be placed below the surface of the ground, so that the top edge of the hopper may be level with the surface, that the peat may be dumped from the car by which it is taken from the bog, and carried to the hopper without hand labor; and this conveyor to be so arranged that the peat will be delivered into the Triturator without hand labor.

Fifth.—CONVEYOR.—Another conveyor, precisely like the one above described, is to be placed so as to convey the peat from the Manipulator into the Tank without hand labor.

Sixth.—TANK.—A tank 35 feet high and 15 feet in diameter; the bottom of this tank is made sloping towards the sides, at an angle of 65°, and is covered with sole tile or drain tile, and the entire inside of this tank is also ribbed with these tile; the ends of these pipes of tile being left open, so that the water which percolates through the pores of the tile, by the pressure of the column of peat, will pass out at the bottom, through the false floor of the tank into the drain, and the solid peat is retained in the tank. A worm is fixed in the bottom of this tank, which is driven by machinery, which forces out the peat in the form of brick, which are cut to any length, and stacked up in sheds, for fuel, after it is fully dried by the air.

*Versmann's Machine.** — This machine, see Fig. 20, was invented by a German engineer, in London, and was patented there in Sept., 1861. It consists of a funnel or hollow cone *b*, of boiler-plate, from one to two feet in diameter at top, and perforated with 200 to 300 small holes per square foot of surface, within which rapidly re-

Fig. 20.—VERSMANN'S PEAT PULVERIZER.

volves an iron cone *a*, carrying on its circumference two spiral knives. The peat thrown in at the top of the funnel is carried down by the knives, and at once cut or broken and forced in a state of fine division through the holes of the funnel, as through a colander. The fine peat collects on the inclined bottom of the chamber *d*, whence it is carried by means of Archimedean screws to a moulding machine. The coarse stuff that escapes pulverization falls through *e* into the cavity *c*. It may be employed as fuel for the engine, or again put through the machine

* Dingler's Journal, Bd. 168, S. 306, und Bd. 172, S. 332.

This machine effects a more perfect pulverization of the peat, than any other hitherto described. This extreme division is, however, unnecessary to the perfection of the product, and is secured at great expense of power. Through the opening at the bottom of the funnel, much unpulverized peat finds its way, which must be continually returned to the machine. Again, stones, entering the funnel, are likely to break or damage the spiral knives, which bear close to the walls of the funnel.

The pulverized peat must be moulded by hand, or by a separate instrument.

*Buckland's Machine** is identical in principle with Versmann's, and in construction differs simply in the fact of the interior cone having spiral grooves instead of spiral knives. This gives greater simplicity and durability to the machine. It appears, however, to require too much power to work it, and can hardly equal other machines in the quantity of product it will deliver for a given expenditure. The ground peat yielded by it, must be moulded by hand, or by other machinery. This machine, we understand, has been tried near Boston, and abandoned as uneconomical.

The machines we have described are by no means all that have been proposed and patented. They include, however, so the author believes, all that have been put into actual operation, at the date of this writing, or that present important peculiarities of construction.

The account that has been given of them will serve to illustrate what mechanism has accomplished hitherto in the manufacture of peat-fuel, and may save the talent of the American inventor from wasting itself on what is already in use, or having been tried, has been found wanting. At present, very considerable attention is devoted to

* Described in Journal of the Society of Arts, 1860, p. 437.

the subject. Scarcely a week passes without placing one or more Peat-mill patents on record. In this treatise our business is with what has been before the public in a more or less practical way, and it would, therefore, be useless to copy the specifications of new, and for the most part untried patents, which can be found in the files of our mechanical Journals.

14. *Artificial Drying of Peat.*

As we have seen, air-dry peat contains 20 to 30 and may easily contain 50 *per cent.* of water, and the best hot-made machine peat contains 15 *per cent.* When peat is used as fuel in ordinary furnaces, this water must be evaporated, and in this process a large amount of heat is consumed, as is well understood. It is calculated, that the temperature which can be produced in perfectly burning full-dried peat, compares with that developed in the combustion of peat containing water, as follows:—

Pyrometric effect of perfectly dry peat	4000° F.
" " peat with 30 *per cent.* of water	3240° "
" " " 50 " "	2848° "

But, furthermore, moist or air-dried peat does not burn in ordinary furnaces, except with considerable waste, as is evident from the smokiness of its flame. When air-dried peat is distilled in a retort, a heavy yellow vapor escapes for some time after the distillation begins, which, obviously, contains much inflammable matter, but which is so mixed and diluted with steam that it will not burn at all, or but imperfectly. It is obvious then, that when a high temperature is to be attained, anhydrous or full-dried peat is vastly superior to that which has simply been cured in the open air.

Notice has already been made of Weber's drying-house, the use of which is an essential part of his system of producing peat-fuel. Various other arrangements have been proposed from time to time, for accomplishing the same object. It appears, however, that in most cases the anticipations regarding their economy have not been fully realized. It is hardly probable, that artificially dried peat can be employed to advantage except where waste heat is utilized in the operation.

A point of the utmost importance in reference to the question of drying peat by artificial warmth is this, viz.: Although the drying may be carried so far as to remove the whole of the water, and produce an absolutely dry fuel, the peat absorbs moisture from the air again on exposure; so that drying to less than 15 *per cent* of water is of no advantage, unless the peat is to be used immediately, or within a few days. The employment of highly dried peat is consequently practicable only for smelting-works, locomotives, and manufacturing establishments, where it may be consumed as fast as it is produced.

A fact likewise to be regarded is, that artificial drying is usually inapplicable to fresh peat. The precautions needful in curing peat have already been detailed. Above all, slow drying is necessary, in order that the blocks shrink uniformly, without cracking and warping in such a way as to seriously injure their solidity and usefulness. In general, peat must be air-dried to a considerable extent before it can be kiln-dried to advantage. If exposed to dry artificial heat, when comparatively moist, a hard crust is formed externally, which greatly hinders subsequent desiccation. At the same time this crust, contracting around the moist interior, becomes so rifted and broken, that the ultimate shrinkage and condensation of the mass is considerably less than it would have been had the drying proceeded more slowly.

Besides Weber's drying oven, the fuel for firing which is derived without cost from the stumps and roots of trees that are abundant on the moor, at Staltach, and which are thus conveniently disposed of, we have briefly to notice several other drying kilns with regard to all of which, however, it must be remarked, that they can only be employed with profit, by the use of waste heat, or, as at Staltach, of fuel that is comparatively worthless for other purposes.

The *Peat Kilns* employed at Lippitzbach, in Carinthia, and at Neustadt, in Hanover, are of the kind shown in

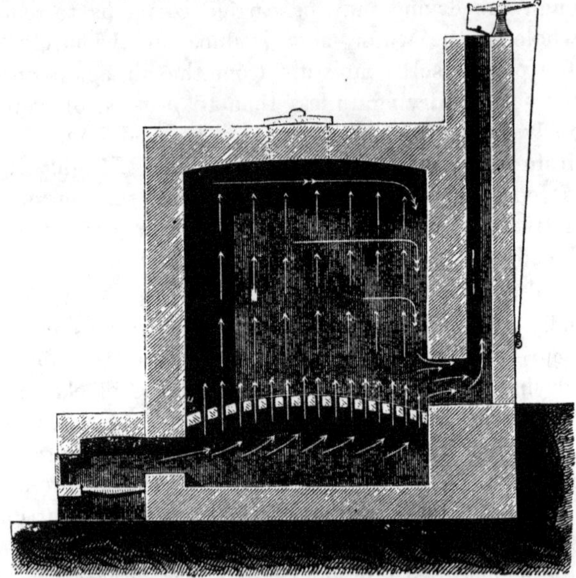

Fig. 21.—CARINTHIAN PEAT DRYING KILN.

fig. 21. The peat with which the main chamber is filled, is heated directly by the hot gases that arise from a fire made in the fire-place at the left. These gases first enter a vault, where they intermingle and cool down somewhat; thence they ascend through the openings of the brick grating, and through the mass of peat to the top of the chamber. On their way they become charged with vapor, and falling, pass off through the chimney, as is indicated by the arrows. The draught is regulated by the damper on the top of the chimney. To manage the fire, so that on the one hand the chimney is sufficiently heated to create a draught, and on the other waste of fuel, or even ignition of the peat itself is prevented, requires some care.

In *Welkner's Peat Kiln** (fig. 22) the peat, previously air-dried, is exposed to a stream of hot air, until it is completely desiccated, and the arrangement is such, that air-dried peat may be thrown in at the top, and the hot-dried fuel be removed at the bottom, continuously.

In the cut, *A* represents the section of a wooden cylinder about 10 feet wide and 6½ feet deep, which surmounts a funnel of iron plate *A'*. The mouth of the funnel is closed by a door *n*; about 20 inches above the door the pipe *B*, which conducts hot air, terminates in the ring *a a*, through the holes in which, *e e*, it is distributed into the funnel filled with peat. The air is driven in by a blower, and is heated by circulating through a system of pipes, which are disposed in the chimney of a steamboiler. From time to time a quantity of dried peat is drawn off into the wagon *D*, which runs on rails, and a similar amount of undried peat is thrown in above.

According to Welkner, a kiln of the dimensions stated, which cost, about $1800 gold, is capable of desiccating daily ten tons of peat with 20 *per cent.* of water, using thereby 2000 cubic feet of air of a temperature of 212° F. When the air is heated by a fire kept up exclusively for that purpose, 10 *per cent.* of the dried peat, or its equivalent, is consumed in the operation. At the Alexis Smelting Works, near Lingen, in Hanover, this peat kiln

* Bornemann & Kerl's Berg und Huettenmænnische Zeitung, 1862, 221.

furnishes about half the fuel for a high furnace, in which bog iron ore is smelted. The drying costs but little, since half the requisite heat is obtained from the waste heat of the furnace itself.

The advantages of this drying kiln are, that it is cheap in construction and working; dries gradually and uni-

Fig. 22.—WELKNER'S PEAT DRYING KILN.

formly; occupies little ground, and runs without intermission.

Other drying ovens are described in Knapp's *Lehrbuch der Chemischen Technologie*, 3. Aufl. Bd. 1, Theil 1, pp. 178–9; *Jahrbuch der Bergakademien Schemnitz* und *Leoben*, 1860, p. 108, 1861, p. 55; Wagner's *Jahres-*

bericht der Chemischen Technologie, 1863, p. 748; Zerrenner's *Metallurgische Gasfeuerung in Oesterreich*; Tunner's *Stabeisen- und Stahlbereitung*, 2. Auflage, Bd. 1, pp. 23–25.

15. *Peat Coal, or Coke.*

When peat is charred, it yields a coal or coke which, being richer in carbon, is capable of giving an intenser heat than peat itself, in the same way that charcoal emits an intenser heat in its combustion than the wood from which it is made.

Peat coal has been and is employed to some extent in metallurgical processes, as a substitute for charcoal, and when properly prepared from good peat, is in no way inferior to the latter; is, in fact, better.

It is only, however, from peat which naturally dries to a hard and dense consistency, or which has been solidified on the principles of Challeton's and Weber's methods, that a coal can be made possessing the firmness necessary for furnace use. Fibrous peat, or that condensed by pressure, as in Exter's, Elsberg's, and the Lithuanian process, yields by coking or charring, a friable coal comparatively unsuited for heating purposes.

A peat which is dense as the result of proper mechanical treatment and slow drying, yields a very homogeneous and compact coal, superior to any wood charcoal, the best qualities weighing nearly twice as much per bushel.

Peat is either charred in pits and heaps, or in kilns. From the regularity of the rectangular blocks into which peat is usually formed, it may be charred more easily in pits than wood, since the blocks admit of closer packing in the heap, and because the peat coal is less inflammable than wood coal. The heaps may likewise be made much smaller than is needful in case of wood, viz.: six to eight feet in diameter, and four feet high. The pit is arranged as follows: The ground is selected and prepared as for charcoal burning, and should be elevated, dry and compact. Three stout poles are firmly driven into the ground, so as to stand vertically and equi-distant from each other, leaving within them a space of six or eight inches. Around these poles the peats are placed endwise, in concentric rows to the required width and height, leaving at the bottom a number of air-channels of the width of one peat, radiating from the centre outwards. The upper layers of peat are narrowed in so as to round off the heap, which is first covered with dry leaves, sods, or moss, over which a layer of soil is thrown. Dry, light wood being placed at the bottom of the central shaft, it is kindled from one of the canals at the bottom, and the charring is conducted as is usual in making wood coal. The yield of coal ranges from 25 to 35 *per cent.* of the peat by weight, and from 30 to 50 *per cent.* by volume.

Gysser recommends to mould the peat for charring in the form of cylinders of 3 to 4 feet long, which, when dry, may be built up into a heap like wood.

A great variety of ovens or kilns have been constructed for coking peat.

At the Gun Factory of Oberndorf, in Wirtemberg, peat is charred in the kiln represented in the accompanying figure. The chamber is 9 feet high, and 5½ feet in diameter. The oven proper, *b b*, is surrounded by a mantle of brick *a a*, and the space between, *c c*, is filled with sand. Each wall, as well as the space, is 15 inches in thickness, and the walls are connected by stones *d d*, at intervals of three feet. Above the sole of the kiln, are three series of air holes, made by imbedding old gun barrels in the walls. The door, which serves to empty the kiln, is a plate of cast iron, the sides of its frame are wider than the thickness of the wall, and by means of a board *e*, a box *m* can be made in front of the door, which is filled with sand to prevent access of air. The peat is filled in through *i*, a channel being arranged across the bottom of the kiln, from the door *f*, for kindling. When the firing begins, the lowest air-holes and *i* are open. When, through the lower gun barrels, the peat is seen to be ignited, these are corked, and those above are opened. When the smoke ceases to escape above, all the openings are closed, *m* is filled with sand, *i* is covered over with it, and the whole is left to cool. It requires about 8 to 9 days to finish the charring of a charge. Several kilns are kept in operation, so that the work proceeds uninterruptedly.

Fig. 23.—OBERNDORFER PEAT CHARRING KILN.

At Staltach, Weber prepares peat coal in a cylinder of sheet iron, which is surrounded by masonry. Below, it rests on a grating of stout wire. Above, it has a cover, that may be raised by a pulley and on one side is attached a small furnace, figure 24, the draught of which is kept up by means of a blower, or an exhaustor, and the flame and hot gases from it, *which contain no excess of oxygen*, play upon the peat and decompose it, expelling its volatile portions without burning or wasting it in the slightest degree. The construction of the furnace, see fig. 24, is such, that the sticks of wood, which are employed for fuel, are supported at their ends on shoulders in the brick-work, and the draught enters the fire above instead of below. The wood is hereby completely consumed, and by regulating the supply of air at *a* (fig. 25) by a sliding cover, and at

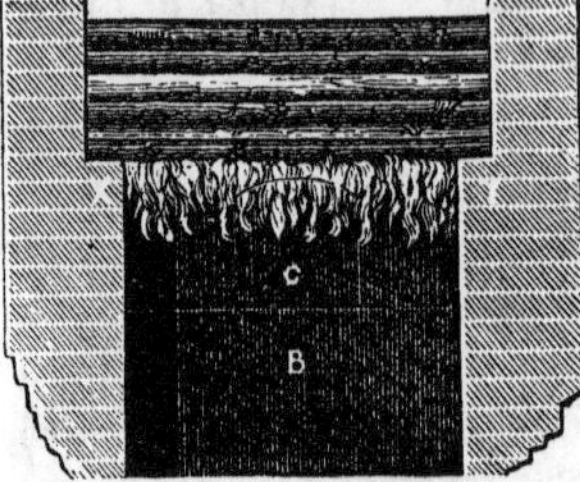

Fig. 24.—WEBER'S CHARRING FURNACE.—TRANSVERSE SECTION.

Fig. 25.—WEBER'S CHARRING FURNACE.—LONGITUDINAL SECTION.

b by a register, the flame and current of air which enters the cylinder containing the peat, is intensely hot and accomplishes a rapid carbonization of the peat, but as before stated, does not burn it. In this furnace the wood, which is cut of uniform length, is itself the grate, since iron would melt or rapidly burn out; and the coals that fall are consumed by the air admitted through *c*. The hot gases which enter the cylinder filled with peat near its top, are distributed by pipes, and, passing off through the grating at the bottom, enter the surrounding brick mantle. Before reaching the exhaustor, however, they pass through a cooler in which a quantity of tar and pyroligneous acid is collected.

Weber's oven is 15 feet in diameter, and 3½ feet high; 528 cubic feet of peat may be coked in it in the space of 15 hours. The wood furnace is 2 feet in section, and consumes for the above amount of peat 3½ cwt. of wood. So perfectly are the contents of the iron cylinder protected from contact of oxygen, that a rabbit placed within it, has been converted into coal without the singeing of a hair; and a bouquet of flowers has been carbonized, perfectly retaining its shape. The yield of coal in Weber's oven is nearly 50 *per cent.* of the peat by weight.

Whenever possible, charring of peat should be carried on, or aided by waste heat, or the heat necessary to coking should be itself economized. In manufacturing and metallurgical establishments, a considerable economy in both the drying and coking may often be effected in this manner.

On the bog of Allen, in Ireland, we have an example of this kind. Peat is placed in iron ovens in the form of truncated pyramids, the bottoms of which consist of movable and perforated iron plates. The ovens are mounted on wheels, and run on a rail track.

Five ovens filled with peat are run into a pit in a drying house, in which blocks of fresh peat are arranged for drying. Each oven is connected with a flue, and fire is applied. The peat burns below, and the heat generated in the coking, warms the air of the drying house. When the escaping smoke becomes transparent, the pit in which the ovens stand is filled with water slightly above their lower edges, whereby access of air to the burning peat is at once cut off. When cool, the ovens are run out and replaced by others filled with peat. Each oven holds about 600 lbs. of peat, and the yield of coal is 25 *per cent.* by weight. The small yield compared with that obtained by Weber's method, is due to the burning of the peat and the coal itself, in the draught of air that passes through the ovens.

The author has carbonized, in an iron retort, specimens of peat prepared by Elsberg's, Leavitt's, and Aschcroft and Betteley's processes. Elsberg's gave 35, the others 37 *per cent.* of coal. The coal from Elsberg's peat was greatly fissured, and could be crushed in the fingers to small fragments. That from the other peats was more firm, and required considerable exertion to break it. All had a decided metallic brilliancy of surface.

16.—*Metallurgical Uses of Peat.*

In Austria, more than any other country, peat has been employed in the manufacture of iron. In Bavaria, Prussia, Wirtemberg, Hanover, and Sweden, and latterly in Great Britain, peat has been put to the same use. The general results of experience, are as follows:—

Peat can only be employed to advantage, when wood and mineral coal are expensive, or of poor quality.

Peat can be used in furnaces adapted for charcoal, but not in those built for mineral coal.

Good air-dry peat, containing 20 to 30 *per cent.* of water, in some cases may replace a share of charcoal in the high furnace.

At Pillersee, in Austria, spathic iron ore has been reduced by a mixture of fir-wood charcoal, and air-dry peat in the proportions of three parts by bulk of the former to one of the latter. The use of peat was found to effect a considerable saving in the outlay for fuel, and enabled the production to be somewhat increased, while the excellence of the iron was in no way impaired. The peat was of the best quality, and was worked and moulded by hand.

When the ore is refractory and contains impurities that must be fluxed and worked off in slag, a large proportion of air-dry peat cannot be used to advantage, because the evaporation of the water in it consumes so much heat, that the requisite temperature is not easily attained.

At Achthal, in Bavaria, air-dry peat was employed in 1860, to replace a portion of the firwood charcoal, which had been used for smelting an impure clay-iron-stone: the latter fuel having become so dear, that peat was resorted to as a make shift. Instead of one "sack," or 33 cubic feet of charcoal, 24 cubic feet of charcoal and 15 cubic feet of peat were employed in each charge, and the quantity of ore had to be diminished thereby, so that the yield of pig was reduced, on the average, by about 17 *per cent.* In this case the quality of the iron, when worked into bar, was injured by the use of peat, obviously from an increase of its content of phosphorus. The exclusive use of air-dry peat as fuel in the high furnace, appears to be out of the question.

At Ransko, in Bohemia, *kiln-dried peat*, nearly altogether free from water, has been employed in a high furnace, mixed with but one-third its bulk of charcoal, and in cupola furnaces for re-melting pig, full-dried peat has been used alone, answering the purpose perfectly.

The most important metallurgical application of peat is in the refining of iron.

Dried peat is extensively used in puddling furnaces, especially in the so-called gas puddling furnaces, in Carinthia, Steyermark, Silesia, Bavaria, Wirtemberg, Sweden, and other parts of Europe. In Steyermark, peat has been thus employed for 25 years.

Air-dry peat is, indeed, also employed, but is not so well adapted for puddling, as its water burns away a notable quantity of iron. It is one of the best known facts in chemistry, that ignited iron is rapidly oxidized in a stream of water-vapor, free hydrogen being at the same time evolved.

In the high furnace, *peat-coal*, when compact and firm (not crumbly) may replace charcoal perfectly, but its cost is usually too great.

When peat or peat-coal is employed in smelting, it must be as free as possible from ash, because the ash usually consists largely of silica, and this must be worked off by flux. If the ash be carbonate of lime, it will, in most cases, serve itself usefully as flux. In hearth puddling, it is important not only that the peat or peat-coal contain little ash, but especially that the ash be as free as possible from sulphates and phosphates, which act so deleteriously on the metal. The notion that, in general, peat and peat charcoal are peculiarly adapted for the iron manufacture, because they are free from sulphur and phosphorus, is extremely erroneous. Not infrequently they contain these bodies in such quantity, as to forbid their use in smelting.

In the gas-puddling furnace, or in the ordinary reverberatory, impure peat may, however, be employed, since the ashes do not come in contact with the metal. The only disadvantage in the use of peat in these furnaces is, that the grates require cleaning more frequently, which interrupts the fire, and, according to Tunner, increases the consumption of fuel 8 to 10 *per cent.*, and diminishes the amount of metal that can be turned out in a given time by the same quantity.

Notwithstanding the interruption of work, it has been found, at Rothburga, in Austria, that by substitution of machine-made and kiln-dried peat for wood in the gas-puddling furnace, a saving of 50 *per cent.* in the cost of bar iron was effected, in 1860. What is to the point, in estimating the economy of peat, is the fact that while 6.2 cubic feet of dry fir-wood were required to produce 100 lbs. of crude bar, this quantity of iron could be puddled with 4.3 cubic feet of peat.

In the gas furnace, a second blast of air is thrown into the flame, effecting its complete combustion; Dellvik asserts, that at Lesjœforss, in Sweden, 100 lbs. of kiln-dried peat are equal to 197 lbs. of kiln-dried wood in heavy forging. In an ordinary fire, the peat would be less effective from the escape of unburned carbon in the smoke.

In other metallurgical and manufacturing operations where flame is required, as well as in those which are not inconvenienced by the ingredients of its ash, it is obvious that peat can be employed when circumstances conspire to render its use economical.

17.—*Peat as a source of illuminating gas.*

Prof. Pettenkofer, of Munich, was the first to succeed in making illuminating gas from wood; and peat, when operated according to his method, furnishes also a gas of good quality, though somewhat inferior to wood-gas in illuminating power.

It is essential, that well-dried peat be employed, and the waste heat from the retorts may serve in part, at least, for the drying.

The retorts must be of a good conducting material; therefore cast iron is better than clay. They are made of the ⌓ form, and must be relatively larger than those used for coal. A retort of two feet width, one foot depth, and 8 to 9 feet length, must receive but 100 lbs. of peat at a charge.

The quantity of gas yielded in a given time, is much greater than from bituminous coal. From retorts of the size just named, 8000 to 9000 cubic feet of gas are delivered in 24 hours. The exit pipes must, therefore, be large, not less than 5 to 6 inches, and the coolers must be much more effective than is needful for coal gas, in order to separate from it the tarry matters.

The number of retorts requisite to furnish a given volume of gas, is much less than in the manufacture from coal. On the other hand, the dimensions of the furnace are considerably greater, because the consumption of fuel must be more rapid, in order to supply the heat, which is carried off by the copious formation of gas.

Gas may be made from peat at a comparatively low temperature, but its illuminating power is then trifling. At a red heat alone can we procure a gas of good quality.

The chief impurity of peat-gas is carbonic acid: this amounts to 25 to 30 *per cent.* of the gas before purification, and if the peat be insufficiently dried, it is considerably more. The quantity of slaked lime that is consumed in purifying, is therefore much greater than is needed for coal-gas, and is an expensive item in the making of peat-gas.

While wood-gas is practically free from sulphur compounds and ammonia, peat-gas may contain them both, especially the latter, in quantity that depends upon the composition of the peat, which, as regards sulphur and nitrogen, is very variable.

Peat-gas is denser than coal-gas, and therefore cannot be burned to advantage except from considerably wider orifices than answer for the latter, and under slight pressure.

The above statements show the absurdity of judging

of the value of peat as a source of gas, by the results of trials made in gas works arranged for bituminous coal.

As to the yield of gas we have the following data, weights and measures being English: —

100 lbs. of peat of medium quality from Munich, gave Reissig	303	cub. ft.
" air-dry peat from Biermoos, Salzburg, gave Riedinger	305	"
" very light fibrous peat, gave Reissig	379 to 430	"
" Exter's machine-peat, from Haspelmoor, gave.............	367	"

Thenius states, that, to produce 1000 English cubic feet of purified peat-gas, in the works at Kempten, Bavaria, there are required in the retorts 292 lbs of peat. To distil this, 138½ lbs. of peat are consumed in the fire; and to purify the gas from carbonic acid, 91½ lbs. of lime are used. In the retorts remain 117 lbs. of peat coal, and nearly 6 lbs. of tar are collected in the operation, besides smaller quantities of acetic acid and ammonia.

According to Stammer, 4 cwt. of dry peat are required for 1000 cubic feet of purified gas.

The quality of the gas is somewhat better than that made from bituminous coal.

18.—*The examination of Peat as to its value for Fuel*, begins with and refers to the air-dry substance, in which:

1.—Water is estimated, by drying the pulverized peat, at 212°, as long as any diminution of weight occurs. Well-dried peat-fuel should not contain more than 20 *per cent.* of water. On the other hand it cannot contain less than 15 *per cent.*, except it has been artificially dried at a high temperature, or kept for a long time in a heated apartment.

2.—*Ash* is estimated by carefully burning the dry residue in 1. In first-rate fuel, it should amount to less than 3 *per cent.* If more than 8 *per cent.*, the peat is thereby rendered of inferior quality, though peat is employed which contains considerably more.

3. — *Sulphur* and *phosphorus* are estimated by processes, which it would be useless to describe here. Only in case of vitriol peats is so much sulphur present, that it is recognizable by the suffocating fumes of sulphuric acid or of sulphurous acid, which escape in the burning. When peat is to be employed for iron manufacture, or under steam boilers, its phosphorus, and especially its sulphur, should be estimated, as they injure the quality of iron when their quantity exceeds a certain small amount, and have a destructive effect on grate-bars and boilers. For common uses it is unnecessary to regard these substances.

4.—The quantity of *coal* or *coke* yielded by peat, is determined by heating a weighed quantity of the peat to redness in an iron retort, or in a large platinum crucible, until gases cease to escape. The neck of the retort is corked, and when the vessel is cool, the coal is removed and weighed. In case a platinum crucible is employed, it should have a tight-fitting cover, and when gases cease to escape, the crucible is quickly cooled by placing it in cold water.

Coal, or coke, includes of course the ash of the peat. This, being variable, should be deducted, and the *ash-free coal* be considered in comparing fuels.

5.—The *density* of peat-fuel may be ascertained by cutting out a block that will admit of accurate measurement, calculating its cubic contents, and comparing its weight with that of an equal bulk of water. To avoid calculation, the block may be made accurately one or several cubic inches in dimensions and weighed. The cubic inch of water at 60° F., weighs 252½ grains.

www.ingramcontent.com/pod-product-compliance
Lightning Source LLC
LaVergne TN
LVHW021359110826
845150LV00007B/1706

* 9 7 8 1 4 2 5 5 1 3 3 9 9 *